TV Repair
for Beginners

THOUSAND OAKS LIBRARY
1401 E. Janss Road
Thousand Oaks, CA 91362

TV Repair
for Beginners

Revised and Expanded

Homer L. Davidson

McGraw-Hill

New York San Francisco Washington, D.C. Auckland Bogotá
Caracas Lisbon London Madrid Mexico City Milan
Montreal New Delhi San Juan Singapore
Sydney Tokyo Toronto

Library of Congress Cataloging-in-Publication Data

Davidson, Homer L.
    TV repair for beginners / Homer L. Davidson.—Revised and Expanded
        p.   cm.
    Previous ed. by George Zwick
    Includes index.
    ISBN 0-07-015805-3 (hardcover).—ISBN 0-07-015806-1 (softcover)
    1. Television—Repairing—Amateurs' manuals.   I. Zwick, George
TV repair for beginners. II. Title.
TK9960.D38   1998
621.388'87—dc21   97-25446
                                       CIP

621.388

McGraw-Hill

A Division of The McGraw-Hill Companies

2 3 4 5 6 7 8 9 0   DOC/DOC 9 0 1 0 9 8 7

ISBN 0-07-015806-1 (HC)
ISBN 0-07-015805-3 (PBK)

The sponsoring editor for this book was Scott Grillo, the editing supervisor was John M. Morriss, and the production supervisor was Clare Stanley. It was set in ITC Century Light by Tanya Howden of McGraw-Hill's desktop publishing unit in Hightstown, N.J.

Printed and bound by R. R. Donnelley & Sons Company.

McGraw-Hill books are available at special quantity discounts to use as premiums and sales promotions, or for use in corporate training programs. For more information, please write to the Director of Special Sales, McGraw-Hill, 11 West 19th Street, New York, NY 10011. Or contact your local bookstore.

EL1

This book is printed on recycled, acid-free paper containing a minimum of 50 percent recycled, de-inked fiber.

Contents

3 Color TV and how it works *69*

4 Color-only sections *97*

8 Troubleshooting catastrophic failures   *207*

Introduction

Besides knowing how the various circuits in the work and why they fail, many people would like to try making a few simple TV repairs. There are many TV repair problems you can make through out the book, changing the fuse, resetting the circuit breaker, cleaning the tuner, connecting the antenna cable, and troubleshooting many different circuits with the digital multimeter.

Most of the service problems throughout the book can be tested with the digital multimeter (DMM) and transistor tester. Those two test instruments can be found on the service bench. If you do not have a DMM, purchase a good portable meter that can measure voltage, resistance, current, capactance, diode test, frequency, and transistor tester. A meter with these functions can be purchased for around $150. Of course, the more test equipment you have and know how to operate, a lot more tests can be made on the TV. Throughly learn how each test instrument works before attempting tests on the TV.

Before tearing into any electronic chassis, you should know the hazzards and saefty factors. Never remove parts or components with the chassis plugged into the ac receptacle. Never take resistance measurements with the power turned on. Pull the ac cord and discharge the main filter capacitor when taking critical resistance measurements in the low-voltage circuits. Likewise, never solder any connection with the power turned on. Only have the power on when taking voltage measurements or waveforms. Remember, the 120-Vac power line can be dangerous when crossed or mistreated. Respect all voltages when servicing.

When taking voltages or waveforms on a live TV, use an isolation transformer to prevent damage to the TV circuits and connected test equipment. Besides internal TV and test equipment damage, you might accidentaly jump and pull test equipment off the service bench and damage the surrounding equipment. You could receive a shock between the test equipment and the TV chassis. Think before making any tests within the TV chassis. Do not work on the TV near metal posts, furnace ducts, and damp or wet areas.

Be careful around the picture tube when the TV chassis power is on. Keep hands and body away from the deflection yoke and the high-voltage anode connection. The

HV can arc up to a few inches—especially on large picture tubes that operate with approximately 42 kV volts. Do not use a multimeter to measure high voltage at the CRT. Not only will you destroy the meter, but receive a terible shock with the test leads. Use only the correct high-voltage test instrument when taking high-voltage measurements. Remember, the picture tube can hold a high-voltage charge for several days after the TV is shut off. Do not be afraid of the TV chassis, just respect it!

Learn the basics of TV servicing, on how to make correct voltage, resistance, transistor, and IC tests inside of the TV. You can test each transistor with the diode-junction test of the DMM. Besides the diode and transistor tests, a few multimeters have transistor HFE tests. Often, low-resistance or continuity tests are indicted on the LCD display with an audible continious beep or tone, indicating a low resistance (under 200 Ω).

Many new TV circuits have appeared on the scene in the last 10 years. Examples are the X-ray protection circuits, scan-derived and secondary flyback voltages, saw filter networks, comb filters, high-voltage shutdown circuits, chopper and switching power supplies, sand-castle generators, stereo sound, on-screen display, surface-mouneted components, and progarm review (sometimes called "picture in a picture"). All of these new circuits are described in this edition. Also, you will find how to install and operate the RCA DSS system, which reaches millions of viewers to receive many differnt channels on high-quality digital video and sound programs anywhere in the USA.

Here are many new symptoms and service problems that you can solve with just two simple test instruments. In fact, there are several tests the beginner can make without any problems. Included in Chapter 2, how to install the RCA DSS system and a wireless telephone system. You can also learn how to replace a defective picture tube with the directions given in Chapter 7. Do not be afraid to call on the professional electronic technician on service problems beyond your own knowledge and when requiring critical test equipment.

Each chapter has been expanded and revised to cover the new TV circuits. Besides adding new circuits and block diagrams, there are over 350 illustrations, with over 150 photos to illustrate the various components found in the latest TV. Learning how the TV operates and fails can be quite rewarding. Making those first repairs can be very satisfying. Besides, electronic servicing is a lot of fun.

The technical level of this book is necessarily different for different users. Although the more experienced reader might find some of the detailed explanations somewhat naive, these same explanations are nothing short of astounding revelations to the uninitiated. However, the overall level of the subject matter is such that the average reader will be able to progress from page to page without difficulty. What might, at first glance, seem too technical and overwhelming will, in time, become fairly simple and relatively easy to accomplish.

TV Repair
for Beginners

1
CHAPTER

The TV system

Before embarking on a detailed, nontechnical description of various portions of a modern TV system, it is desirable to present a brief synopsis of the overall process, from end to end. This can provide a continuity of visualization so that as you read about a particular step in the system, you will have in the back of your mind, so to speak, the complete picture of the overall purpose and end result to be achieved.

Basic sound and picture systems

The TV system, whether monochrome (black-and-white) or color, consists of two distinct transmitter-receiver systems: the sound (audio) system and the picture (video) system. The interconnections and any similarities between the two systems are entirely incidental. In other words, it is not essential to have any common connections or functions between the two systems. It is done merely for economy and convenience. A physically separate sound system alongside a corresponding totally independent picture system would be just as feasible from a strictly technical viewpoint.

The sound system is essentially the same as the FM portion in AM/FM receivers. The FM sound part of the TV signal picked up by your antenna has the same general characteristics of an FM signal—high immunity to noise (static, electrical interference, etc.), relatively short-range reception capability, and (potentially) a higher sound quality. Figure 1-1 shows, in block diagram form, the major components of the TV sound system.

Figure 1-2 shows, in similar form, the building blocks of the video (picture) system. This is an AM signal (similar to the AM music stations) and is subject to the same general conditions accompanying AM station reception—fading, freak long-distance reception, interference (streaks, etc., across the screen), and so on.

Terminology

Amplitude modulation (AM) signifies a system of radio transmission and reception in which the magnitude of the signal varies directly with the loudness of the voice or music or with the brightness of the image, in the case of a TV picture.

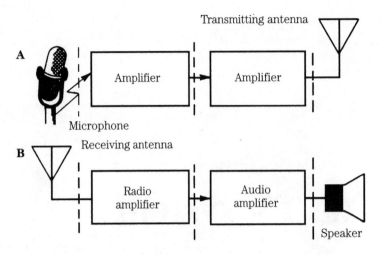

1-1 A simplified block diagram of a sound transmitter (A) and a sound receiver (B). Notice that the two systems are virtually mirror images of each other.

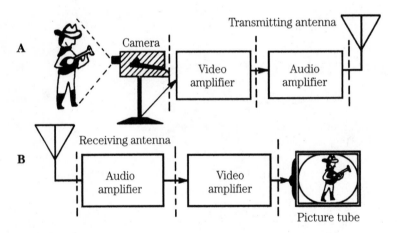

1-2 A simplified block diagram of a video transmitter (A) and a video receiver (B).

Because electrical noise can add to or subtract from such a signal, it is only logical that such noise is reproduced by the receiver as physical noise (static, clicks, etc.) or as streaks, dashes, etc., on the TV screen. In other words, AM is not immune to noise.

Frequency modulation (FM) ignores any variations in size of the radio signal as might be caused by the addition of noise to the signal, and transmits and receives sound information by means of a variation in the spacings (modulation) between adjacent waves. Technically, both picture and sound could be AM or FM, but there are other reasons for the present use of FM sound and AM picture.

Transmitting and receiving systems

In Figs. 1-1A and 1-2A, the vertical dashed lines define the three distinct major functions of each transmitting system. On the extreme left of Fig. 1-1A is a microphone, which changes sound waves to electric currents. In Fig. 1-2A, the camera performs the corresponding function of changing light into electric energy.

The second sections of Figs. 1-1A and 1-2A serve the major functions of amplification (enlargement) of the faint signals. The third sections in both illustrations convert the amplified signals into a form suitable for transmission via the antennas. They also further amplify them before feeding them to the antennas. The right-hand sections in both illustrations actually radiate the signals into space to be received by many receiving antennas (Fig. 1-3).

TV receiving antennas are available in many different sizes and shapes. For long-distance signals, the yagi antenna might have 10 to 13 directors for the low channels, and 10 or more directors for the high VHF band (Channels 2 through 13). The UHF antenna might have a separate yagi or corner reflector to cover channels 14 on upward. Some yagi antennas have the UHF antenna mounted on the front end of the VHF antenna and connected to the VHF matching harness. In the normal

1-3 The VHF yagi type antenna receives channels 2 through 13. Notice that the small end of the antenna points toward the station being received.

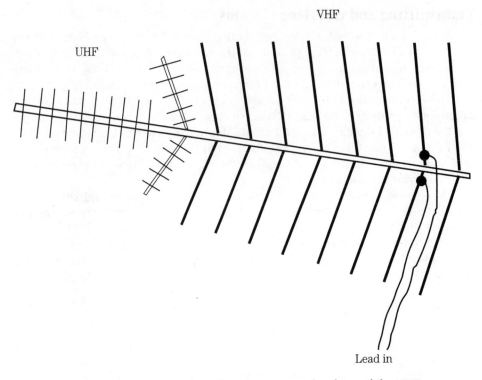

VHF

UHF

Lead in

1-4 The UHF antenna with a corner reflector is mounted in front of the VHF antenna.

VHF antenna, the directors, folded-dipole, and reflector rod are at the rear of the complete antenna mast (Fig. 1-4).

The UHF antenna can have a corner reflector at the back of the yagi antenna. You might find 10 to 15 short dipoles as directors, with a folded-dipole next to the corner reflector. The shorter the dipole or rods, the higher the frequency received. The TV signal is picked up by the folded-dipole and matched to the transmission line. The more elements used as directors, the greater "pulling power" of the antenna. The Ultra High Frequency (UHF) signal does not travel out very far and should have a good transmission line of either 300-Ω foam-filled line or shielded 75-Ω cable to the TV set for best reception.

Original sound to reproduced sound

All sound (voice, music, or noise) is produced by a physical force that sets the air into vibration. The amount of force applied to the air determines the loudness of the sound, and the different pitch (from bass to a whistle) results from the rate (speed) of vibration of the air.

Microphone

The microphone is a device that converts mechanical energy (air pressure or vibration) into electrical energy. The amount of the electrical energy produced depends on the physical force applied (sound loudness). The pitch of the original sound determines the frequency (numerical rate of vibration) of the electrical energy produced. For sound, the frequency might be as low as 20 vibrations per second (a deep, organ note) and as high as 20,000 vibrations a second (a very high-pitched whistle). Incidentally, ordinary household ac has a frequency of 60 vibrations (cycles or hertz) per second.

Amplifier

The amplifier is an electrical device, using either tubes or transistors, that has one function—to faithfully enlarge the feeble electrical currents emanating from the microphone. No one tube or transistor can provide all the amplification that is required, so there usually are a number of amplifiers, each amplifying the output of the preceding amplifier or stage until the required maximum is obtained (Fig. 1-5).

Audio and radio frequencies

To attain a useful understanding of TV transmission and reception, it is essential to have a clear concept of frequency. As was mentioned earlier, electrical energy exists as a wave-like phenomenon, just as sound energy consists of air waves, not unlike the waves or ripples caused by a disturbance in water. The term *frequency* is used to denote the number of times a wave recurs in a certain period of time, usually one second. Thus, in all future references to *frequency* in this book, the meaning will

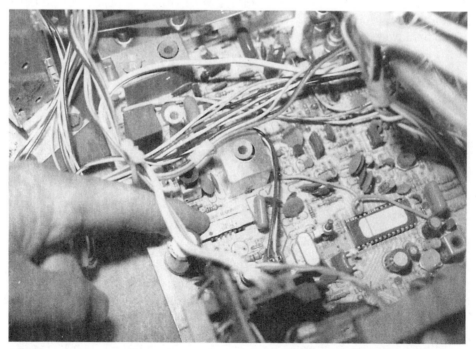

1-5 The solid-state audio section might contain one large IC with sound IF and output circuits.

be the same: the number of waves per second (preferably, cycles per second). As stated before, audio frequencies (AF, sound frequencies) are assumed to extend from about 20 cycles per second (deep organ note) to about 20,000 cycles per second (for a very shrill whistle). Above 20,000 or 30,000 cycles, they are called *radio frequencies (RF)*, extending through AM and FM radio, TV, and radar transmitting and receiving frequencies to the near-optical (light) portion of the spectrum. Here are the first infrared (invisible heat waves), followed by the visible light spectrum (red through violet), the ultraviolet (again invisible), x-rays, etc. Video (picture) frequencies are related to television. They represent for those frequencies that carry picture information and extend from about 20 cycles per second to about 4,000,000 cycles per second.

Until several decades ago, frequencies in the entire spectrum from audio through invisible light were referred to as *cycles per second, thousands of cycles per second (kilocycles, kcs), millions of cycles per second (megacycles, mcs)*, etc. Now by international standards, the term *hertz (Hz)* is used instead of cycles per second. Heinrich Hertz, a German physicist, was the first to demonstrate the production and reception of electromagnetic or radio waves. Thousands of cycles is referred to as *kilohertz (kHz)*, etc.

A final note on waves and frequencies: the higher the frequency, the shorter the length of the wave. For example, the sound wave of a telephone bell (with a frequency of, for example, 1000 hertz) is approximately 186 miles long. By contrast, the radio wave transmitted by a UHF television station is little more than 1 foot long. This explains why the elements or metal rods of a UHF TV antenna are so much shorter than the corresponding VHF types. Although any frequency can be transmitted from place to place over wires, only those classed as radio frequencies (RF) can be sent through space (wireless transmission).

RF and AF amplifiers

The middle sections of the simple diagrams in Figs. 1-1 and 1-2 consist mainly of amplifiers, accessory devices, and controls. Those immediately following the microphone are audio amplifiers, as stated previously; however, to transmit this audio economically and efficiently over great distances a temporary change is made. The audio wave is superimposed (that is, it modulates) on a radio wave piggyback style. This combined RF and AF energy is further amplified and finally applied to the antenna (Fig. 1-6).

Antennas

A transmitting antenna has one function: to radiate the energy applied to it in the desired direction (or in all directions) in the most efficient manner possible. Although the obvious purposes of the receiving and transmitting antennas seem to be quite different from each other, they actually behave very much alike. Incidentally, despite many claims to the contrary, there is no substitute or shortcut worthy of its name for a high-grade, elaborate antenna, whether for transmitting or receiving.

At the receiving end, the antenna does not radiate energy—it collects. Located in the path of the radiated energy, the receiving antenna collects or acquires a very small sampling of the original energy or signal as radio waves cross it. By means of a

1-6 The IF and video section of the TV front end is usually shielded.

transmission line (sometimes called *lead-in*), the intercepted signal is carried to the input of the receiver (Fig. 1-7).

Although a properly designed receiving antenna will favor the stations (frequencies) for which it was designed and discriminate against others, it is fairly helpless in rejecting undesirable electrical energy, such as noise, static, etc. The transmission line is equally susceptible to noise or static pickup, although something can be done here. Shielding, protecting the lead-in from surrounding electrical noise, is feasible. The antenna itself cannot be shielded (or it won't receive any signals); the only recourse lies in the selection of a suitable location where noise is at a minimum.

Booster systems

TV signals that must be picked up a greater distance, (50 to 100 miles), can be helped with a booster system. A high-gain antenna amplifier system can deliver a 25-dB VHF and 20-dB UHF signal for better reception. Some booster systems have a top-side booster that mounts upon the antenna mast close to the antenna, and the bottom half is mounted in the basement or on the rear of the TV receiver. The top-half booster amplifies the TV signal and pushes the strong signal down the shielded or foam-filled lead in cable to the bottom half of the booster unit. Usually, the bottom half contains the dc power supply and the termination antenna terminals (Fig. 1-8).

Large high-gain indoor antenna boosters can be located in the basement, with several different TV set hookups. The booster and antenna terminals are contained in one unit. Because of the weak signal and matching cables to several TV receivers in

1-7 The two top antennas receive channels 2 through 13, with a corner reflector for channel 17 and 21, and a UHF flat-bed antenna to receive the local UHF station.

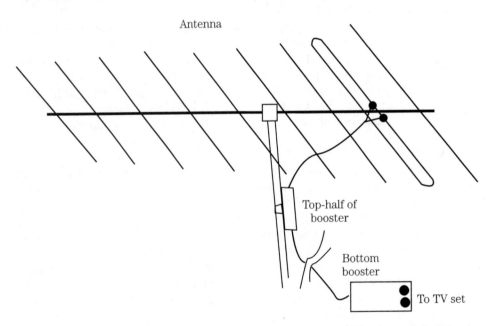

Antenna

Top-half of
booster

Bottom
booster

To TV set

1-8 One half of the top-side booster mounts on the mast and the bottom half in the basement or on the rear of the TV set.

the house, the TV signal must first be amplified. The 300- or 75-Ω coax cable distribution amp can drive four or more TV sets. This reduces the UHF/VHF/FM signal loss caused by sharing the signal among several TV sets or long antenna runs (Fig. 1- 9). The antenna booster plugs directly into the power line outlet at all times.

RCA DSS system

The RCA DSS system (Digital-Satellite-System) is a direct broadcast from a satellite system that enables millions of people to receive over 100 channels of high-quality digital signals from anywhere in the USA. The system provides digital data, video, and audio to the various homes via a high powered KU-band satellite. The broadcast signal is transmitted to the DBS satellites orbiting the earth above 22,000 miles, above the equator, from Colorado Springs, Colorado. This signal is relayed back to earth and decoded via a receiver.

Two different packages are available. The model DS1120RW is a basic package with antenna dish and a single-output, low-noise block converter (LNB) (Fig. 1-10). This satellite receiver has a DSS/TV universal remote.

The deluxe package contains a Sheet Molded Compound (SMC) antenna with a twin dual-polarity LNB. The satellite receiver has an output so that two different TVs can be plugged in with additional audio/video jacks. A fully universal remote is included that can also operate the satellite receiver and VCR.

The antenna dish has an 18-inch diameter with slight oval shape for the KU-band. The LNR is offset so that it is out of the way and does not block any surface of the dish (Fig. 1-11). The low-noise block converter converts the 12-GHz to 12.7-GHz signal from the satellites to a 950-MHz signal that can be used by the receiver. The dual-output LNR

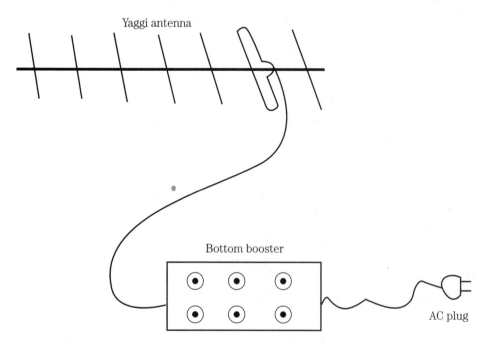

1-9 The high gain booster can be located in the basement with several TV jacks for additional TV hookups.

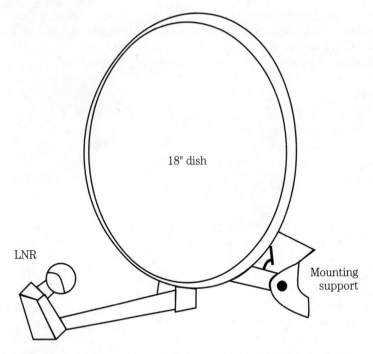

18" dish

LNR

Mounting
support

1-10 The RCA DSS 18-inch dish antenna can be mounted anywhere, so long as the line between the antenna and satellite is
clear.

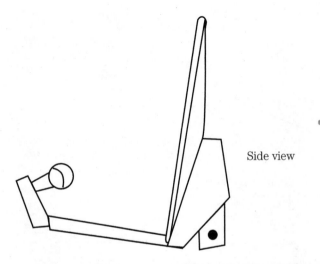

Side view

1-11 The side view of a slightly oval dish with
mounting brackets and angle division lock assembly.

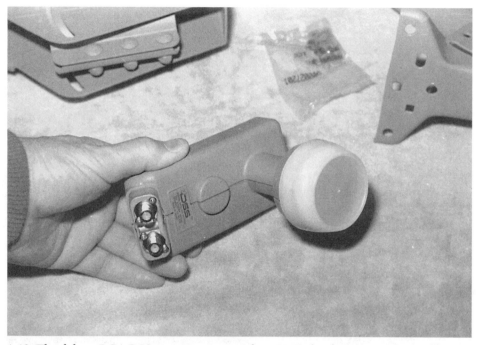

1-12 The deluxe RCA DSS system comes with an LNR that has two outlets and can be connected to two different satellite receivers.

is used in the deluxe package. The dual-output LNR can be used to feed the second receiver or another form of distribution system (Fig. 1-12).

The basic package costs around $499, while the deluxe DSS system sells for $699 (and these prices are dropping). These prices are quite low compared to the first satellite systems that were introduced above $10,000. This DSS system has regular satellite and cable systems beat in cost, as well as the excellent digital video and sound. The DSS system can be utilized in remote locations that can not hookup to cable TV.

Receiver

The amplifier sequence in the receiver is virtually a mirror reflection of that in the transmitter. Amplifiers immediately following the antenna amplify the combined (RF plus AF or modulated RF) signal to the required level. Next, and in the same general portion of the receiver, a reversal of the modulation process occurs via a device called a *demodulator* or *detector*. This reversal consists of stripping the RF, which has served its purpose, from the signal, leaving only the audio-frequency signal. Further amplification now occurs until the AF signal is sufficiently strong to drive a speaker or headphones (Fig. 1-13).

Speaker

In function, a speaker or loudspeaker is the exact opposite of the microphone. Electrical energy in the form of audio waves is applied to the speaker, where it is

1-13 Some TVs are still manufactured with manual-rotary tuners, including a separate VHF and UHF tuners.

converted to the mechanical motion of a surface called the *speaker cone*. The cone vibrates faster or slower, depending on the frequency (pitch) of the corresponding original sound waves, and sets the air in similar motion. The loudness of the sound depends on the distance that the cone moves, which, in turn, depends on the strength or size of the electrical wave applied to it (Fig. 1-14).

Original picture image to radio signal

In the following paragraphs, I trace the sequence of steps necessary to convert a visual scene into an electromagnetic energy wave suitable for transmission in space. After the conversion from light to electrical energy has been accomplished in the camera, the remaining functions are very similar to the sound-to-RF-to-sound process just described. The functions to be described apply equally to monochrome and color transmission reception and should give you a clear nontechnical concept of what is going on so that you will be able to deduce what went wrong and where in the event your TV set performs unsatisfactorily or fails to perform altogether.

Physical image

All images—persons, scenes, or whatever—can be seen by the human eye because of the light they emit or reflect. It is quite obvious that all such images are color images. In addition, the various colors reaching the eye are of different light intensity: dark brown, or dazzling white, or bright yellow, etc. The best example of this is a so-called black-and-white picture made by a camera, where all colors become different shades of gray. A person's face is brighter than his or her shoes, but not as bright as a white shirt, etc. In a manner of speaking, our descriptions of different colors are given in terms of brightness or dimness, with the notion of color being a separate characteristic of the object or scene we are viewing.

Photographic recording

To the human eye, a scene or object presents a continuous range of color and illumination. Actually, however, this seeming continuous image is a composite of many discrete little areas, each of which might differ from its neighbor either in color, brightness, or both. In a black-and-white photograph, the image is actually made up of tiny specks of the image spaced from each other. The concept of *grain* (fine-grain photograph, for instance) refers to just this particle structure. Although the color photograph is claimed to be grainless, this is a relative term, in contrast with the chemical grain structure of the black-and-white picture.

Halftone

A practical example of a discrete particles composition of an image is the familiar newspaper or magazine photograph. Viewed with the naked eye, such a photograph looks smooth and continuous. Under a magnifying glass, however, the picture is revealed as consisting of individual dots with clear spaces between dots. Shading in the picture depends on the weight of the black dots in comparison to the adjacent

1-14 Large 10- or 12-inch speakers are sometimes used in console TVs. Oval speakers are often used in today's portable TVs.

white spaces. Although this composition results from the technique of mass reproduction (newspaper printing, etc.), images of very high quality can be and are reproduced by the discrete dot structure.

Conversion of image dots into electric signals

The first step in producing a television signal is the conversion of light (optical energy) into electricity. It begins with a photographic-type camera, lenses, focusing, viewfinders, etc.; however, here is where the similarity ends. Instead of projecting the image onto a ground glass (as in studio cameras) or on a photographic film (any camera), it is projected onto a special electrochemical surface called a *photomosaic*. Incidentally, the projected image might be quite small, often not much larger than a postage stamp.

Next, an electron beam or stream scans the mosaic image, one spot at a time, as if reading the bits of the image. Each time the electron stream impinges on one of the spots of the mosaic image, an electric current is generated that has characteristics particular to that particle. That is, the electric pulse or current from that image particle contains information necessary to reproduce this image particle on the home TV screen.

Also included with each such electric pulse is the position or location information, both in the vertical and the horizontal directions. This positioning information is transmitted and received by the home TV antenna, faithfully amplified and reproduced, and finally used to guide the electron beam in the TV picture tube for proper positioning of the image particle.

A simple example of an electron gun used to produce a scanning beam is in the neck portion of a TV picture tube. It is composed of a glowing heater—a sort of electron generator—and a number of metallic structures (hollow cylinders, baffles, etc.) that shape and guide the electrons toward the screen in the form of an extremely thin, sharp beam. In the home receiver, the electron beam "writes" on the picture tube screen in step with the "reading" performed by the beam in the TV camera. The location and brightness of each written bit is automatically known from information transmitted with each bit so that the completed written image is an exact duplicate of the original. The scanning sequence is arranged in accordance with an established code (National Television Standards Committee, NTSC). In simplified form, it is as illustrated in Fig. 1-15. (This scanning sequence can be considered, for the sake of illustration, analogous to the writing of a letter, double-spaced, then filling in the spaces between the lines so that the complete letter is read line after line, from top to bottom.)

Beginning at the upper-left corner, the electron beam scans line 1 from left to right. It then skips back to the left side and scans the second line underneath the first. This is line 2. This sequence continues until approximately 235 lines have been produced. In Fig. 1-3, all these lines are shown solid. The beam continues to scan an additional $27\frac{1}{2}$ lines, for a total of $262\frac{1}{2}$, with but one difference—these additional lines are not visible on the home screen—their location is just below the visible bottom of the picture. These "invisible" lines are explained later. The beam now skips back to the upper-left corner and starts scanning again, but this time between the lines just finished, producing lines 263, 264, etc., until about line 498, then continues

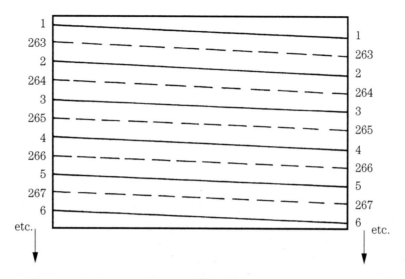

1-15 How a TV picture is scanned by electronic beams. The solid lines represent one field, and the dashed lines show the alternate field. The typical sequence is 1, 2, 3,...up to 262 lines. A complete screen continues to a total of 525 scanning lines in the NTSC video standard (used in North America).

scanning invisible lines again until 525 are produced. The complete process now repeats, starting with line 1.

The picture tube screen has now been scanned twice, each scan lasting $\frac{1}{60}$ of a second, the period of which is called a *field*. The two fields, lasting a total of $\frac{1}{30}$ of a second, are called a *frame*. The process of alternate line/field scanning is called *interlacing*. In case of a still scene, more than one frame of the same scene would be scanned; in action scenes, at least one frame (two fields) is always scanned, which is required for a correct picture of the scene.

High-definition TV (HDTV)

High-definition TV is a new system that improves picture resolution, with twice the horizontal and vertical resolution. Besides picture definition, it has improved audio, somewhat like the compact disc. The new system does not interfere with existing transmission of the present-day television channels set by National Television Standards Committee (NTSC) of 30 years ago. Your present TV will not be obsolete.

The FCC provides standards that require transmission systems to operate within specific widths and frequencies used in present-day VHF and UHF TV bands. Within these parameters, there are gaps where inactive channels exist. HDTV will use these inactive channels and at the same time broadcast over existing broadcast channels. To enjoy the benefits of HDTV, you must purchase a new TV with HDTV features.

The present-day standard TV receiver has 262.5 scanning lines in the odd and even field scanning lines. The interpolated-definition TV (IDTV) receivers introduced by foreign manufacturers has 525 odd and even field scanning lines. The interpolated memory digital circuits provide additional scanning lines between the broadcast signal, showing the entire image upon the TV screen (Fig. 1-16).

Several American and Japanese firms plan to manufacture HDTV receivers by 1995. Zenith Electronics and American Telephone & Telegraph Co. are working on a $24 million research and development system of the HDTV system. On December of 1989, the Duran-Leonard fight was televised in high-definition television.

Blanking and synchronizing

In describing the scanning sequence it was stated that at the end of each line, the beam skips back to the beginning of the next line, and at the end of a field, it skips to the upper-left corner of the beginning of the next field. The very important activity during the blanked interval, blanking, simply means making the lines invisible because they are not part of the image and would only add meaningless lines, bars, and streaks to the picture. This activity consists of transmission of synchronizing pulses, which are guidelines for starting and ending of each line. These pulses occur at the end of each line, while the timing of each field is sent during the blank period between fields. Any deficiency or degradation of these sync pulses would be cause for picture instability, such as horizontal tearing or vertical rolling. Other signals are sent during the blanking period, such as color correction (VIR) in the most recent expensive TV sets (referred to later in the book).

In addition to the brightness and contrast controls on the home TV set that are adjustable, some definite normal levels of brightness and contrast are preset and identified by the transmitted signal, incident to the blanking interval information. Thus, it is possible for the home TV viewer, by misadjusting the brightness control while at the same time misadjusting the vertical hold control so that the picture begins to roll slightly, to actually see the vertical blanking and synchronizing pulses on the screen, located between adjacent picture frames. Of course, such excessive brightness setting is detrimental to normal picture light distribution, causing light areas to be burned out with loss of detail. It similarly affects the opposite end of the illumination range, preventing black areas from being actually black. This applies equally to color as to black-and-white sets; thus, indirectly, the appearance of any such blanking lines on the screen is a sign of excessive brightness in some TV sets, assuming that the set is normal otherwise.

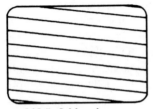

262.5: Odd and even field $\frac{1}{60}$-second scanning lines in present-day standard TVs

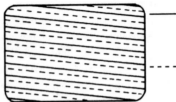

Regular broadcast signal

Interpolated (IDTV) digital memory signal

525: Odd and even field $\frac{1}{60}$-second scanning lines in IDTV

1-16 The 262.5 odd and even field with 525 odd and even fields with an interpolated (IDTV) digital memory signal between.

Interlacing and pairing

If an ordinary photograph were to be cut into a large number of very thin strips, the original image could still be seen, no worse than before, provided that the individual strips were kept from moving, sliding, or shifting. If, however, some or all of the strips were moved so that they partially or completely overlap, the picture would naturally suffer, depending on the degree of overlap. With regard to the TV picture, it is quite correct to visualize it as being made up of such very narrow strips (called *lines*) arranged in two steps. In step one, every other strip is temporarily omitted. This produces a picture with a number of gaps. In step two, these gaps are filled with the remaining set of strips—each exactly between the two adjacent strips. The picture is now complete.

If, however, the second set of strips were placed so that they partially or even totally overlap the first set, the picture will not only be incomplete because some of the strips are covered over, but also because there is a mixing or blurring of whatever is still visible. This phenomenon of overlapping and blurring is called *pairing* because there no longer are any single, individual lines, but pairs of such lines. As explained later, such pairing is caused by a failure of the horizontal sync—either because of a defect in the TV set or because of transmission difficulties, making each line double-width and resulting in half of the normal number on the screen.

Automatic brightness control

In recent years, manufacturers have added a number of so-called "automatic" features to their receivers, usually for the sake of simplification of operation by the user. Sometimes, however, such a feature is a protective measure against improper adjustment and possible consequent damage to the set. Such is the case with ABC (automatic brightness control) or ABL (automatic brightness limiter). From a technical viewpoint, this feature protects some internal parts—especially the picture tube, from premature failure from excessive cathode current; it also keeps the image on the screen from becoming extremely overbright because of careless setting of the front-panel brightness control. As a consequence, the previously mentioned trick to display the blanking and sync bar might not work without a bit of trying and careful misadjustment. It is assumed that after such a trick, the set owner will return the controls to normal by first reducing them to below normal, then gradually bringing them up to normal.

Conversion of electrical impulses into radio waves

Electrical impulses that represent individual image elements and timing (synchronizing) pulses are combined with or superimposed upon a radio wave (carrier), amplified to the power level required for transmission over the intended distance (a TV station is licensed for a certain maximum power level to meet the needs of the geographic area it is to serve) and applied to the antenna.

Incidentally, the frequencies of TV stations in the United States are much higher than those used for ordinary AM radio broadcasts for technical reasons (number of stations to be accommodated, shorter distances covered, etc.). Although ordinary AM stations are centered around 1 MHz (megahertz), TV frequencies range from just over 50 MHz to almost 900 MHz. This affects not only the location of TV stations in the radio spectrum, but also poses different requirements on design, complexity, etc.

TV sound

One additional signal is required for a complete TV transmission: the sound (or noise). As stated before, TV sound is transmitted in the FM mode for certain technical reasons, as well as for some advantages (noise immunity, for instance). TV sound could be transmitted over a completely separate system from microphone to loudspeaker. This, however, would be unwieldy, more expensive, and technically less satisfactory. For these and other reasons, TV sound is sent as part of the overall station transmission with many functions, circuits, and actual components shared by the picture and the sound portions of the receiver.

Figure 1-17 shows, in simplified form, the sequence of functions and their combination into a single final signal for transmission. Notice that the sound part of the system resembles the ordinary radio (AM) mentioned previously; however, instead of being applied to the antenna system directly, the sound signal (an FM system) is combined with the TV picture signal. Then both are amplified as an integrated signal and applied to the transmitting antenna. For all practical purposes, this is now a single radio signal containing all the information necessary to reproduce a TV picture, both sight and sound.

Audio processing

In the typical mono audio system, the video signal passes through a 4.5-MHz bandpass filter that separates the video from the sound. One large IC processor contains the limiter, detector, attenuator, and audio out. Inside the processor, the output of the limiter passes on to the sound detector. Here, the tuned LC network recovers the audio signal.

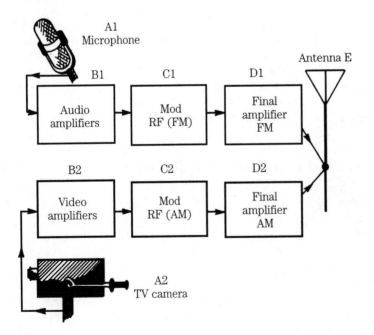

1-17 A simplified block diagram of a TV transmitter from the microphone and TV camera.

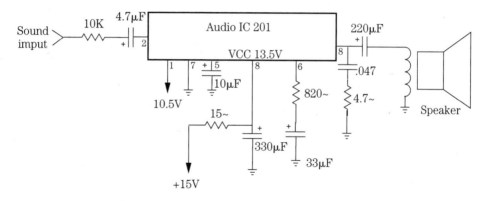

1-18 A simple diagram of the audio mono sound IC circuits.

The attenuator circuit (controlled by the volume control) and audio signal is passed on to the audio AF circuit. In most receivers, the audio is coupled to a pre-amp audio transistor or IC. Some TV audio circuits consist of push-pull audio amp transistors (Fig. 1-18).

Stereo processing

Within the typical stereo audio system, the stereo demodulator contains mono and stereo audio signals. A baseband audio signal is developed and applied to the stereo module. The output signal of IC1 is amplified and passed on to the matrix and differential amp. The left and right signals are applied to the logic monitor IC processor from the matrixing transistors (Fig. 1-19).

The monitor logic IC contains input from the matrix circuits and also has auxiliary audio stereo input jacks. IC3 switches the mono and stereo audio signals. Now, the stereo audio output signal is fed to the volume-control attenuation section. The controlled volume is applied to the audio power amplifier output IC. IC5 provides stereo signal to both stereo speakers. Often, IC processor components are used throughout most stereo processing audio circuits.

Complete TV signal

To summarize the characteristics of a typical TV broadcast signal, as received by the home TV set, the TV signal is a composite of two independent signals or stations—an FM sound transmission and an AM picture transmission. The FM transmission contains just the sound that accompanies the TV picture. The AM transmission consists of the following:

- Picture brightness information, bit by bit.
- Picture background illumination, the overall scene.
- Picture bit location information in the form of cueing signals for starting and stopping each line, field, and frame.
- Picture repetition rate timing signals, which ensure exactly the same line, field, and frame frequency at both the transmitter and receiver.

The third and fourth item of the list are, in effect, two components of control information serving the same general purpose: to ensure that the reproduction of the

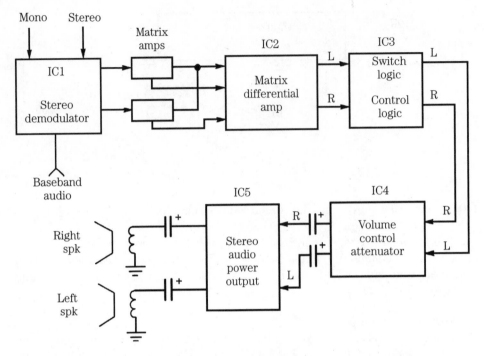

1-19 One of the stereo audio-processing circuits used in today's stereo TVs.

picture on the home TV set is precisely in step with the production of the original picture in the pickup camera. During the description of a TV receiver in Chapter 2, reference is made to the various components of the timing signals, particularly in connection with the subject of vertical hold and horizontal hold.

2
CHAPTER

How a TV receiver works

Some people look at a wiring diagram or a schematic of a TV and see a mass of meaningless lines and symbols, but to an experienced TV servicer, every line and every symbol is significant. They help the servicer follow the path of each signal through each section of the receiver to determine why the set won't work as it should. A schematic also tells the technician what should happen to the signals as they pass through each section. To help you get some idea what's going on in those circuits, consider each part of the TV receiving system (Fig. 2-1).

2-1 A schematic makes TV servicing much easier to locate and check voltages in the TV chassis.

Antenna system

A home TV antenna differs from that used by a TV station for two major reasons. The receiving antenna must be receptive to a large number of stations, each of a different frequency. In other words, the receiving antenna is a broadband device; therefore, some compromise is necessary in comparison to an antenna designed for one particular station only. The latter is the case with the transmitting antenna of any one station. The second difference stems from the fact that the transmitting antenna is intended to radiate in all directions (it is omnidirectional), but the home antenna (even if it has to be aimed in different directions to receive stations from different geographic locations) is, in fact, a highly directional system. Its multidirectional receiving capability is provided by rotating it to the desired position. All of the elements (rods) of the antenna point in the same direction for maximum signal pickup from that direction (Fig. 2-2).

The transmission line or lead-in wire might be a flat-ribbon type or a shielded coaxial cable. Ideally, a flat 300-Ω lead-in wire matches the impedance of the antenna and TV receiver terminals (Fig. 2-3). The 75-Ω shielded coaxial cable must have a 75- to 300-Ω matching transformer at each end of the transmission line for a correct impedance match.

2-2 The VHF yagi antenna is pointed with the small end toward the TV station.

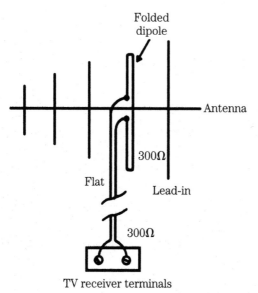

Folded dipole

Antenna

300Ω

Flat

Lead-in

300Ω

TV receiver terminals

2-3
The transmission line can be a 300-Ω flat ribbon cable that matches the TV antenna's and receiver's 300-Ω input.

Shielded cable is an excellent choice when poor TV reception is caused by standing waves (stationary wave distribution of current or voltage) or in areas where excessive noise is generated. Using shielded cable keeps noise and spurious signals from FM, police, and CB transmissions from entering the lead-in wire (Fig. 2-4). Long runs of shielded cable can cause excessive attenuation of the broadcast signal. A booster system (RF amplifier) could solve the problem in weak fringe areas or where two or more TV receivers are sharing the same transmission line. Low-quality TV reception could also result from poor installation or improper location of the transmission line. Problems of and solutions to using inferior transmission lines are considered in the section on troubleshooting.

Common antennas

The yagi antenna is one of the most common directional antennas on towers and rooftops today. A yagi antenna is constructed from rods of various lengths that are spaced in a straight line along a central boom (Fig. 2-5). The rods are the antenna elements. Elements that do not directly receive energy from the transmission line are called *parasitic elements*. The element connected for the efficient transfer of energy between it and the transmission line is called a *driven element*. In a yagi antenna system, a folded dipole is the driven element connected to lead-in wire. The parasitic elements that are shorter than the driven element are called *directors*. They contribute to the antenna's directional characteristic by improving signal reception from the direction that they are pointing. Directors are located on the front end of the boom. Remember that if you ever install this type of antenna, point the end of the boom with the shortest element in the direction of the desired signal

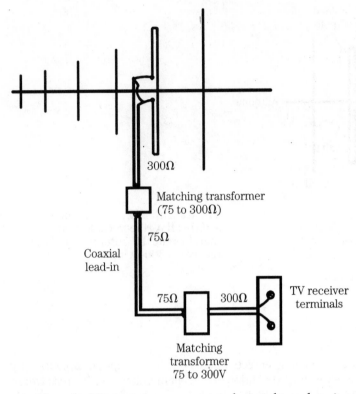

300Ω

Matching transformer
(75 to 300Ω)

75Ω

Coaxial
lead-in

75Ω   300Ω

TV receiver
terminals

Matching
transformer
75 to 300V

2-4 The shielding prevents unwanted signals and noise from entering a long run of 75-Ω coaxial cable.

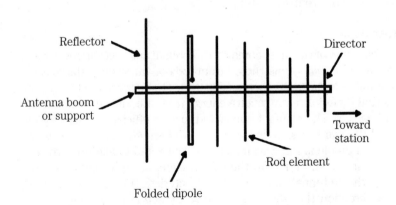

Reflector

Director

Antenna boom
or support

Toward
station

Rod element

Folded dipole

2-5 The yagi antenna has several directors and a long reflector for the most efficient TV antenna.

source. The longest element located at the opposite end or rear of the antenna is called a *reflector*. It aids in the antenna's directional characteristic by canceling or reducing signal reception from the antenna's rear. It is not uncommon to lose at least half of the desired signal strength if this parasitic element is missing or damaged.

Transmission line

The transmission line's sole purpose is to transmit the energy picked up by the antenna to the TV set input with a minimum of loss and, perhaps just as important, with no signal pickup along the way. Because any such pickup is likely to be almost all noise and hardly any signal, the importance of a proper high-quality transmission line cannot be overemphasized.

TV receiver

Figure 2-6 is a block diagram of a TV with each of the major functions of a particular circuit. Now look at each individual stage and see how they work together in the TV chassis. The antenna brings the TV signal into the front-end tuner, which can be controlled with a keyboard or remote transmitter.

The IF cable from the tuner is controlled directly to the video IF/AGC/SIF/Video IC (Fig. 2-7). In the latest TV circuits, IC101 (SIF/Video/Chroma/Vertical/Horizontal) contains most of the TV's circuits. The sound take-off connects to pin 23 to another sound IF and output IC201. Video is taken from pin 18 to the video amps and onto the luminance and chroma circuits. The color output transistors feed the color-gun elements of the picture tube.

The deflection IC supplies a vertical pulse to the vertical output IC and on to the vertical winding of the yoke assembly (Fig. 2-8). Likewise, a horizontal drive pulse or waveform from the deflection IC provides horizontal drive to the driver transistor and transformer coupled to the horizontal output circuits. The collector of the horizontal output transistor connects to the primary winding of the flyback or horizontal output transformer, supplying high voltage, screen, focus, and secondary voltages to the many different circuits.

The low voltage ac power supply contains bridge rectification and feeds a high dc voltage to the IC regulator (Fig. 2-9). IC801 provides a dc voltage to the horizontal driver and output circuits. In some chassis, a voltage divider lowers the voltage fed directly to the deflection IC circuits. With this method, the horizontal and vertical circuits come on at once. The horizontal deflection circuits does not depend on a scan-derived or secondary voltage. In other circuits, the scan- derived or secondary voltage sources must supply voltage to the horizontal deflection IC circuits. If the horizontal circuits are not functioning, all other circuits are dead or in chassis shutdown.

Front end

Boxes A1 and A2 represent devices called *tuners* or *front ends*. Their purpose is to select the desired station and amplify the relatively feeble signal from the antenna. For the models of the past few years, there are two tuners in each TV set: one VHF (very high frequency), covering channels 2 through 13, and UHF

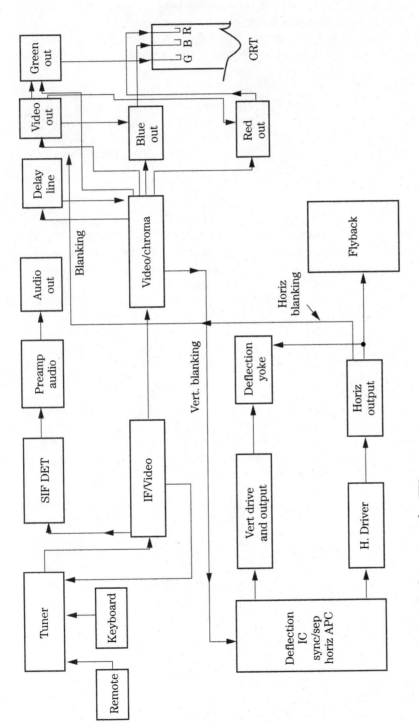

2-6 A block diagram of a present-day TV.

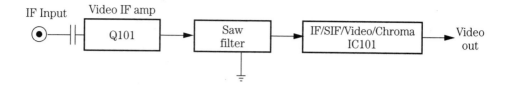

2-7 The IF cable from the VHF and UHF tuner connects to the IF/AGC/SIF/video IC.

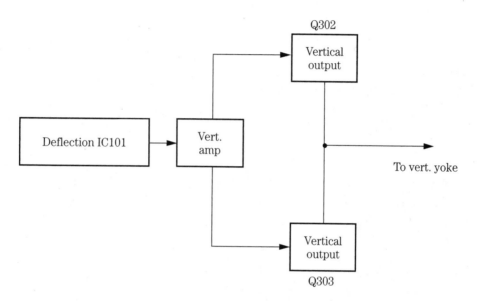

2-8 The deflection IC provides a vertical drive pulse to the vertical output transistors or IC.

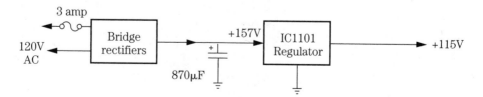

2-9 A block diagram of the low-voltage power supply.

(ultrahigh frequency), covering channels 14 through 83. Both tuners are complex, rather sophisticated assemblies, requiring test equipment and technical know-how that is well beyond the capability of not only the beginner, but even a good segment of the TV repair industry. In fact, many high-grade TV repair shops have TV tuners repaired by special service stations that are fully equipped to do the work. The intent here is to caution the beginner against any rash action in attempting to correct a malfunction in this portion of the receiver. There are, however, certain tasks that are within the ability of the beginner that can be performed satisfactorily, and these are described in chapters covering troubleshooting.

Construction and location of the tuner are of particular interest, more so than any other section of the set (except perhaps the high-voltage cage, which is described later). Both the station selector and the fine-tuning adjustment are, for technical reasons, physically built into the tuners. This requires that the tuner be located at an accessible position on the cabinet. For this reason, as well as for some strictly technical factors, the tuners are physically separate units that are relatively easy to disconnect from the remainder of the TV set (Fig. 2-10).

Interconnections between the tuners and the main TV chassis and the antenna are as follows:

- The signal input from the antenna terminals to the tuners is via twinlead or sometimes coaxial cable.

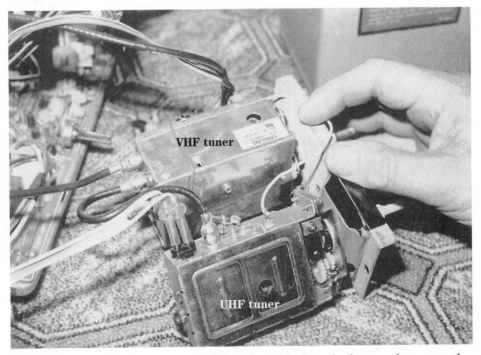

2-10 VHF and UHF tuners are still used in black-and-white and some color portable receivers.

2-11 The tuner can be cleaned with tuner wash or spray if TV channels seem erratic.

- The signal output, almost universally in the form of a shielded coaxial cable (a center wire surrounded by insulation with a braided copper outer sleeve) terminated in a single contact plug.
- A three- or four-wire cable connecting the various voltages in the main chassis. These wires might end in individual lugs fastened by screws, or in a common, multipin plug that fits a mating socket on the main chassis.

Finally, there must be some provision for rigidly fastening the complete tuner assembly to the cabinet. Figure 2-11 is one example of such a tuner; others differ in a minor detail, but essentially they are of the same general construction and interconnection arrangement.

The semiconductor tuner

The change from tube to solid-state TV receivers was made in two steps. The first was to the hybrid (tubes and semiconductors) receiver, and the second to the all-semiconductor set. Strictly speaking, there are no 100 percent solid-state receivers because at least the picture tube is an electron tube. But these sets are commonly described as all solid state. The tuners in all TV sets are of the solid-state variety. The reasons are many and varied, but as far as the nontechnical consumer is concerned, this feature is all the better. These tuners are more stable, more rugged, and longer lasting. Figure 2-12 is a view of a modern solid-state tuner. Greater compactness and freedom from protruding tubes is obvious—even if the durability and greater freedom from repairs might not be visually evident.

2-12 The top tuner is for VHF and the bottom for UHF.

Varactor tuner

A radical departure from conventional tuners, whether tube type or all solid state, is the so-called varactor (voltage-controlled) type. It has many advantages over the conventional type—in stability, freedom from noise, minimal degradation with use, not to mention greater simplicity in construction and maintenance. For an understanding of this, however, it is necessary to have a little background information on tuned circuits in general, and TV tuners in particular.

The term *tuning*, whether applied to automotive or musical instrument adjustments, refers to setting or adjusting a device or a group of devices to a predetermined condition. In electronics, tuning is used exclusively to indicate adjustment of one or two components in a radio receiver (station tuning) or TV set (channel selection). In repair procedures, a function corresponding to tuning is called *alignment*. In each of these examples, tuning is accomplished via a coil (a loop or spool of wire) and a capacitor. In the case of TV tuners, each channel has its own little coil (actually a number of such coils) cut to size and connected to the circuit by means of a multiple-contact switch actuated by turning the channel selector. Because the exact configuration of such a coil or coils is essential for repeatable accurate tuning (getting the correct station reception every time), these coils are made as sturdy and as rigid as possible. Nevertheless, long use, shock, vibration, and switching cause a gradual deterioration of the whole tuning system, eventually requiring replacement of the tuner, or rebuilding—a tedious and expensive job at best. But the weakest component is the switch, operated by the channel-selector knob. The switch contacts, being part of the functioning coil-capacitor assembly, cause the worst degradation with use resulting from contact corrosion, accumulation of dirt and oily film, etc. To the user, this evidences itself in the form of a noisy tuner—erratic pictures, frequent need to jiggle the knob to restore the picture, and ultimately the complete loss of stability (the station will not stay on channel). The varactor tuner goes a long way toward elimination of most of these problems (Fig. 2-13).

In simple terms, the varactor is a semiconductor (a diode, in fact) that not only acts as a capacitor (hence, it is a tuning device), but one whose capacitance can be changed over a considerable range through the application of a voltage. The device is sometimes referred to as *Varicap*™, meaning a variable capacitor. The advantages in TV tuners are obvious. Although station switching by means of the channel selector is still required, switching can now be done in less-critical points, making for much greater stability. Simplification of tuner design and construction is a further advantage of this new technique in semiconductor application to modern TV receivers. Although the advantage might not be obvious to the average user, especially in a new or fairly new receiver, it will become apparent later, as evidenced by the absence of chronic misbehavior of the tuner, as is the case with the coil-switching type.

2-13 The varactor tuner mounts directly on the PC board and contains surface-mounted parts.

In today's TV chassis, you might find that the VHF and UHF tuners are found upon the same PC board. The tuner parts are laid out in a square and must be signal traced and repaired as any electronic component on the PC board. These tuner components are built right on the PC board with surface-mounted parts.

Poor or open grounds within this type tuner can cause many different problems and symptoms in different TV sections. A poor ground or defective tuner component can make the vertical bounce, bunch up, or go into an intermittent horizontal line. The audio might stop, become intermittent and distorted with a defective PC board tuner component or ground. The best method is to solder all tuner grounds on the fine PC wiring. Always use a low-wattage, fine-pointed iron tip.

Tuner module The varactor solid-state tuner can operate in a manual, push-button keyboard, or in remote-control mode. The tuner module is easily exchanged by simply unplugging the wires and the IF cable to the TV chassis, then remove two or more mounting screws to free the tuner. Varactor tuners of the same type are used in other configurations with the push-button or remote-control TV receiver (Fig. 2-14).

Push-button keyboard With keyboard TV operation, simply push the correct numbers to select the desired TV station. To select channel 13, push buttons 1 and 3. Likewise, the UHF station channel 27 can be selected by pushing buttons 2 and 7 (in that order). All VHF and UHF channels are directly accessed, regardless of frequency band or numerical designation (Fig. 2-15).

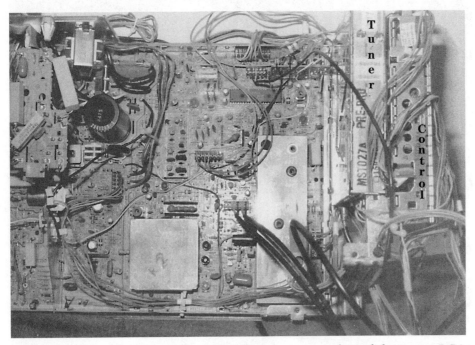

2-14 The VHF and UHF tuner modules with a control module in an RCA CTC131 chassis.

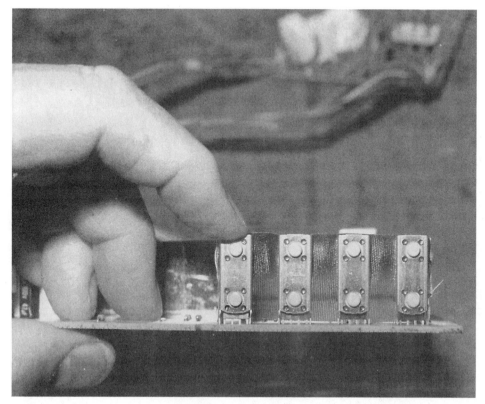

2-15 Push-button tuning in the RCA CTC146 13-inch portable.

Besides having a tuner module, the push-button operation could have a separate synthesis or memory module. Usually, these modules are plugged into each other and receive their operating power from the TV chassis. The antenna cable and IF receiver cable plug into the tuner module. These modules can be easily replaced by removing the plugs and metal mounting screws. Manufacturers might call the various modules by different names, but the operation is the same.

Remote-control tuner modules Since cable TV has become so popular, TV manufacturers have designed the TV antenna terminals to accept the cable connection. The cable system can be connected directly to the tuner module through a 75-Ω cable connector, or a standard 300-Ω antenna connection can be used. Today, more remote-control TV receivers are sold than any other type because all of the stations can be easily selected at random.

In older remote-control TVs, noisy and cumbersome motors were used to rotate the tuner and TV controls. With the new tuner, memory and control modules, the TV receiver can silently be tuned to any TV channel electronically within seconds (Fig. 2-16). The remote tuning is accomplished with a small hand-held transmitter. A receiver is in the control module located within the TV cabinet.

Remote transmitter The present day simple hand-held transmitter can select any channel, turn the receiver off and on, adjust volume up and down, and when the

2-16 The remote control and combination UHF/VHF tuners used in a recent TV.

telephone rings, the mute button can be used to silence the TV (Fig. 2-17). A more-elaborate remote transmitter might have the capabilities to allow quick scanning of the entire TV frequency band within seconds of individual channel selection by pushing the correct channel designation. Besides controlling the TV, it might also control such components as a video disc player and VCR.

One older type of remote transmitter radiates a unique RF signal for each operation mode, and when more than one TV of the same type is operating in the home, the remote-control transmitter could operate other TV sets. This problem could even occur in an apartment complex between neighbors. Today, the infrared transmitter will only operate the TV receiver if the remote transmitter is pointed at the TV. These infrared transmitters are very directional and radiate a short range infrared signal.

The portable hand-held remote transmitter can operate from two small penlight cells or a regular 9-volt transistor battery. The batteries will normally operate satisfactorily for six months to one year; when replacement is required, heavy-duty leakproof types are recommended. Weak batteries might cause the remote hand-held transmitter not to operate at all, or weaken the radiated infrared signal, making it impossible to operate the TV when the transmitter is held more than a foot or two away.

Check the remote transmitter batteries any time that the TV set fails to respond to the infrared signal; try cleaning the battery terminals by rubbing each terminal with a cloth. If the remote control will only operate after tapping on the case, check for loose batteries or a broken connection. The hand-held remote transmitter is often dropped, dislodging components and cracking the PC board.

Remote-control transmitters are often sent to the tuner repair depots that service TV tuner assemblies.

Universal remote transmitter

Universal remotes can operate just about any TV, VCR, CD audio, and video product. Almost all electronic TV manufactures have one. The General Electric RRC500 does the work of three remotes and the RRC600 does the work of four different remotes. A Radio Shack three-in-one remote control can manage TV, VCR, and cable operations (Fig. 2-18).

Unlike some universal remote controls, the three-in-one is preprogrammed and does not have to learn its commands from the original remote control. You just tell the remote which remote control you wish to replace by entering three-digit codes, and the remote does the rest. Another bonus is that the three-in-one universal remote has large numbers and letters to push with white buttons. They are easy to see, and you can easily push the right buttons the first time.

The operation manual shows how to tune in or lock the TV, VCR, or cable into the present TV set. For instance, if you have an RCA TV and know its code, simply start with the various numbers listed in the chart. The numbers given for an

2-17 The universal remote-control transmitter can control TV, VCR, cable, and audio products.

2-18 The Radio Shack three-in-one control has large, easy-to-see buttons to control the TV, VCR, and cable operations.

RCA TV set are, 018, 019, 038, 047, and 135. Start with the first three numbers and continue each set of numbers until the TV turns on and operates with the three-in-one remote control. By pressing the right keys to enter the manufacturer's code, once the code is set, you can now operate the remote at any time. The VCR can be coded in the very same manner.

Universal remotes seem to be more powerful in operation than regular remotes. If the remote becomes weak or only operates part of the time, replace the batteries. Have the new battery replacements handy to insert, before removing any batteries. The three-in-one memory only lasts about one second without the batteries, so change the batteries as quickly as possible. If the memory is lost, simply reenter the three-digit codes.

Frequency-synthesis tuning system

The most expensive and latest RCA chassis have a channel-lock frequency-synthesis tuning system, which can provide up to 127 channels. Also, the tuner covers all of the UHF band. The tuning system can be controlled manually or remotely with a hand-held remote transmitter. A cable/normal switch allows the viewer to decide between regular TV broadcast or cable reception.

The frequency-synthesis board might consist of a synthesizer IC, which contains a PLL step generator, band switch decoder, AFT, digital sync, presence detector, serial decoder, aux switch, and video blanking. The microcomputer section provides correct IR decoding, keyboard decoder, sync communicator clock, and data signal for the synthesizer IC. It also provides a display driver to the channel information buffer. The synthesis tuner might be contained in a module.

Manual preset tuner

The manual preset tuner is operated manually with stations already tuned into a varactor tuner. The varactor tuner tunes in the TV broadcast stations by applying different voltages to a varactor diode within the tuner. The voltage applied must be varied by a variable rheostat control or with preset resistance. In this tuner, a small variable-resistance control is provided, one for each station, including six on the UHF bands (Fig. 2-19).

Because there are only 12 channels, a small variable resistor is included at the front of the tuner for adjustments. Slip off the plastic plate to get at these controls or remove tuner. These adjustments can be made from the front of the TV set. The numbers A through F can be removed and the UHF number placed in each section. Because most areas have only a few UHF stations, the rest of the letters will remain. The six large variable resistors at the bottom represent the UHF stations. If one of

2-19 RCA's preset manual tuner has a variable resistance for each TV station.

the stations slides off or cannot be tuned manually, remove the plate and tune the correct variable control with channel number.

Installing RCA's DSS antenna system

First, select a mounting site for the dish that is clear of all buildings and trees between antenna and satellites in the sky. Remember, the DSS satellites are located over the equator at 101 degrees west longitude. With the azimuth and elevation known, find the correct angle with a compass, level, and protractor to approximately locate the satellite and the angle of the antenna. In the Midwest, the angle for the antenna dish to the satellite signals is about 36 degrees, slightly turned west. In other words, the 18-inch dish would be pointing south and a little west at a 36-degree angle, pointed upward toward the sky (Fig. 2-20).

The DSS dish has two different positioning adjustments, azimuth and elevation. The azimuth adjustment is the side movement of the dish. You can make this adjustment by rotating the dish on the mounting or post. The elevation adjustment is done with the LNB support arm. Notice to the side of this arm, the dish and arm can be adjusted up or down. A reference scale on the side of the dish is calibrated in degrees. Adjust the LNB arm and dish for the correct degree of elevation.

To adjust the dish for elevation, loosen two screw nuts securing the dish. Now the dish will move up and down. Line up the elevation indicator with the elevation degree angle. Tighten the two nuts securing the dish elevation movement.

Point the dish in the general direction of the satellites. Follow the instructions with the service literature on how to make the correct azimuth adjustments with a compass and azimuth coordinates. A fine tuning of the dish position can be made with a signal strength menu and receiver audio tone indicator. This menu is tuned in on the TV screen.

The DSS receiver signal strength screen uses two systems to help fine tune the location of the dish. Listen to the audio tone by connecting a TV, headphones, or an

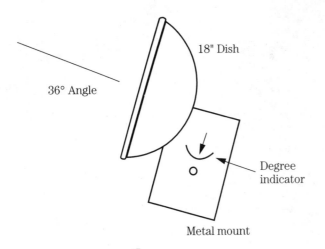

18" Dish

36° Angle

Degree indicator

Metal mount

2-20 The angle setting of the 18-inch dish has an elevation scale on the side of antenna mount.

amplifier to the correct jacks on the rear of the receiver. When the dish is pointed at the satellites, a continuous tone can be heard. When not pointed at the signal, no tone or only short bursts of tone can be heard.

The second tuning system uses the television on-screen menu display. This display includes both a bar and numeric display. The numeric display is from 0 to 100. The higher the number, the greater the signal. Zero is a very weak signal. A signal around 40 to 50 might fade in and out, the picture stripes with picture in a picture, the sound cuts in and out, and the picture goes out completely. The stronger the signal with the bar display, the farther the bar stretches across the screen. By listening to the audio tone and looking at the display, you can fine tune the position of the antenna. A signal above 72 is good and a white bar signal (around 100) is excellent.

The antenna dish arrives with 100 feet of RG-6 coaxial cable and 20 feet of telephone wire. The shorter the transmission line, the greater the reception. Route the cable through the shortest distance. A telephone connection should be installed next to the DSS receiver. The telephone is used for communications.

Mounting the dish

The dish can be mounted just about anywhere, as long as it is free of all objects between satellite and dish. Level the mast holding the antenna with a bubble level or plumb bob. A metal foot mount is located at the base of the mast (Fig. 2-21). Do not mount the foot mount over vinyl or metal siding. When mounting it on lap wood

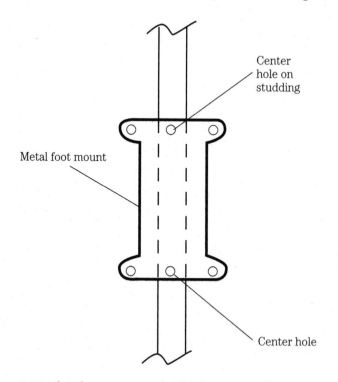

Center hole on studding

Metal foot mount

Center hole

2-21 The foot mount should be centered over a wooden studding with an antenna side mount.

siding, be sure that the center part of the foot mount is over a wooden studding. Do not mount the foot mount on the chip board fiber and particle board.

When mounting the foot mount upon a flat or slanted roof, be sure that the mount straddles a wood rafter. The dish can be mounted into the ground by digging a 2½-foot deep hole that is 10 inches in diameter. Use a 1¼-inch galvanized pipe and pour concrete into the hole. Be sure the mast or pipe is plumb vertically, in all directions. Pound an 8-foot rod nearby and clamp a ground wire to the antenna mast (Fig. 2-22).

The antenna dish can be mounted on a chimney with a chimney-mounting kit. It's best to use a chimney that is not active. Again, the foot mount must be mounted solid at the chimney mount so that the antenna has no objects between it and the satellite.

Installing the LNB cable

If you have installed the deluxe RCA DSS system, three different cables must be installed. Two RG-6 (75 Ω) cables must be attached to the LNB at the dish and run to each receiver on the two different TV sets. The third cable is a telephone wire for each receiver. Only two different cables are needed when installing the basic DSS system. The cable from the dish carries the TV signal from LNB to receiver and a separate telephone wire (Fig. 2-23).

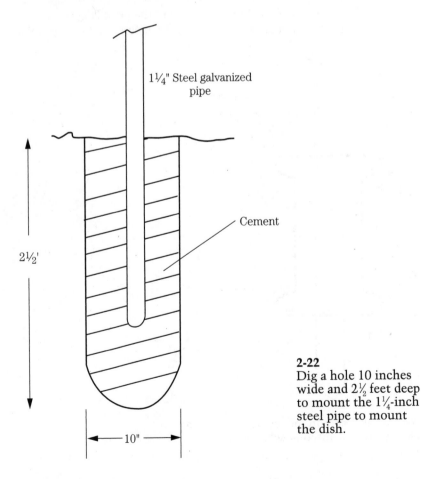

1¼" Steel galvanized pipe

Cement

2½'

10"

2-22
Dig a hole 10 inches wide and 2½ feet deep to mount the 1¼-inch steel pipe to mount the dish.

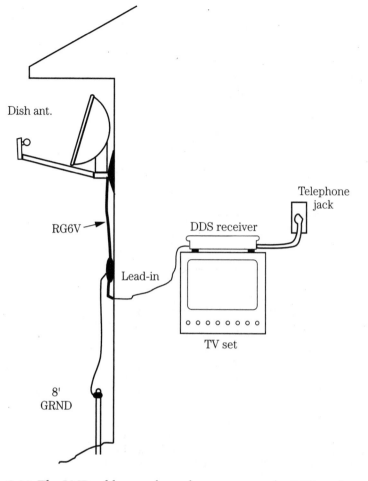

Dish ant.

RG6V

Lead-in

8'
GRND

Telephone
jack

DDS receiver

TV set

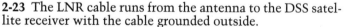

2-23 The LNR cable runs from the antenna to the DSS satellite receiver with the cable grounded outside.

The coaxial cable should be grounded outside where the cable enters the home. The best method to ground the cable is with a ground block. The metal ground block has two cable-barrel coax connections and a ground wire connection. Attach a ground wire to the metal bar and to a common ground stake or water pipe nearby.

The deluxe receiver (DRD102RW) has a satellite jack to plug the coaxial cable into from the antenna dish. If a local antenna or cable antenna system is used, plug it into the correct antenna input jack. The TV output connection goes between satellite receiver and TV set. If a VCR is also connected to the TV set, connect the TV output cable to the input jack of the VCR. Then connect the output jack of the VCR into the TV receiver antenna connection (Fig. 2-24). The telephone jack on the satellite receiver is connected to the telephone line.

Wireless telephone jack system

You might have a very remote location where the telephone wire and jack cannot be installed for the new DSS antenna system. This DSS system requires a

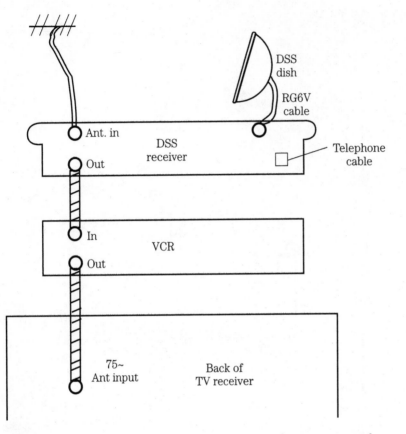

2-24 The LNR RG-6 coaxial cable connects to the receiver with a telephone jack mounted below. Connect the regular TV and cable wires to the in and out receiver terminals on the DSS receiver to the VCR and TV.

telephone connection be made to the DSS receiver for direct communications. Here, a wireless phone jack system fills the bill of no more cutting and drilling into walls, no waste of time in awaiting for the telephone installation. Simply install the wireless phone jack system (Fig. 2-25).

The wireless telephone system contains a base and extension jack unit. The base unit plugs into the existing electrical outlet near the present telephone jack or installation. The telephone cable connects to the base unit. The extension jack unit is also plugged into a wall ac outlet and the telephone connection goes directly to the telephone jack on the DSS receiver. You can also have an extension at this point, if needed (Fig. 2-26).

The wireless telephone system can be used in the house, townhouse, condos, apartments, and home offices. The wireless telephone jack system does the job when its impossible or too costly to run the telephone extension between telephone or jack and satellite receiver. You can also connect another telephone at this location.

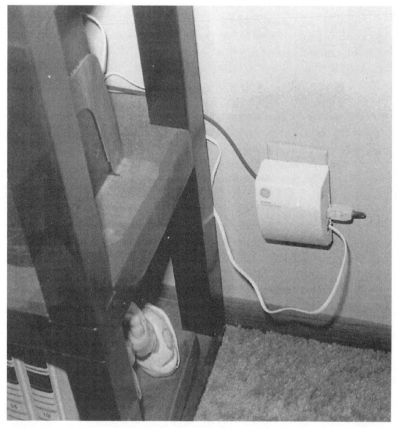

2-25 The wireless telephone jack system can be mounted in difficult places, where no outside telephone jack can be located.

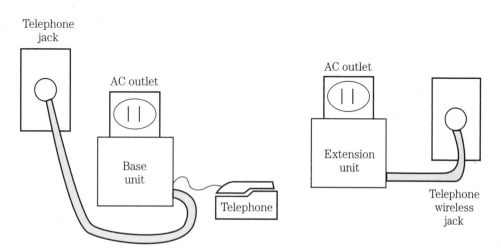

2-26 The telephone base unit plugs into an ac outlet near the telephone and extension unit, where a telephone outlet is needed.

IF amplifiers

Two or three tubes were used in the TV IF stages. Three or more transistors were used in the early solid-state chassis. Today, the whole IF, SIF, and first video stages might be located in one large IC that includes chroma, vertical, and horizontal deflection with drive circuits (Fig. 2-27).

Ahead of the IF IC circuits is a saw filter network and video IF amp transistor. While in the early solid-state IF IC circuits, one IC might receive the signal from the IF cable and fed through an RF coil directly into the IF IC. Besides the IF circuits, AGC circuits can be contained in the same IC. The video output was transformer-coupled to a picture IF transistor, which separates the picture and sound at the collector terminal.

The present-day IC chip might provide IF sound coils and the audio take-off to the audio output IC. A video signal is fed from the IC chip to a 4.5-MHz filter circuit to prevent sound from entering the picture and to one or two video amp transistors. The video circuits are entered into the same large chip and fed to the luminance and color circuits.

There are usually three separate amplifiers or stages in sequence, as links in a chain. A defect or total failure in any one stage breaks the chain and prevents the signal from continuing on its way. There is usually some feedthrough of signal even when one stage goes dead, as in a case of tube or transistor burnout or other failure, so that some small portion of the normal signal is transferred to the next stage. One symptom of such a failure might be a very weak picture on any station; although some stations may come in better than others largely because of the original difference in signal strength between stations.

Symptoms of snow mentioned in regard to the front end are much less apparent here and may even be completely absent. Another result of such a failure might be picture instability, such as tearing of the picture in a generally horizontal (actually diagonal) direction and possibly a rolling picture (vertical direction). This is due to the fact that the amplifiers in block B also must pass the synchronizing pulses. A failure

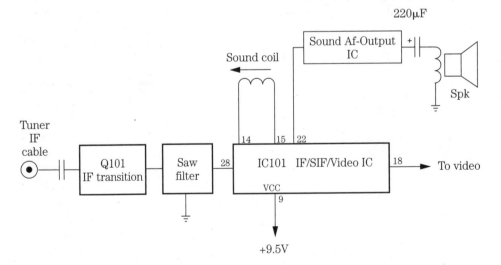

2-27 The PIX IF, sound IF, and video circuits are contained in one large IC in today's TV.

in these amplifiers will invariably degrade the quality of these pulses to a level below that required for picture holding. In the section on troubleshooting, I shall present concrete procedures designed to help locate trouble in the IF portion of the set.

At this point it is imperative to clearly indicate what the beginner may not and should not attempt as far as the IF transformer adjustments. You may see a half-dozen either rectangular metal cans or uncased spool-like coils or transformers, some of them with a hole or slot that seems to invite a screwdriver blade for turning. These are, in fact, adjustments, but they cannot be properly done by eye or ear. Required is some very sophisticated equipment and very specific knowledge and expertise, different for each and every adjustment in all TV sets. Worse yet, unlike the adjustment of some circuits recommended later in this book that can be either repeated or restored to their original position, the adjustment of any of the IF coils is, from the very beginning, a point-of-no return case. Not only is it virtually impossible to know what the adjustment is accomplishing, but it is equally impossible to go back to the starting point.

To further illustrate the futility of such adjustments, the professional service uses complex equipment that gives a visual presentation of the operation of all the circuits at the same time. Only on such a visual display can an expert servicer observe the effect of each adjustment on the overall picture. Looking at the front of the TV set while twiddling one of the adjustments is almost certain to destroy the normal quality of the picture without showing any immediate change as the twiddling is made. Finally, it is fortunate that these adjustments, except for some very minor effects, are virtually permanent and do not require any correction, except by the professional servicer after replacing a transformer in case of a burnout or other catastrophic failure. To sum it up, the best advice is leave the IF adjustments alone. They are probably as they should be.

Transistorized SAW filter and PIX IF

The IF signal is applied to the PIX IF and surface acoustic wave (SAW) filter component coming from the tuner system. Usually, the IF signal is amplified, then applied to the SAW filter network (Fig. 2-28). The output of the SAW filter is applied to the IC processor.

The IC processor might have two or three stages of IF amplification, video detector, AFT, and AGC circuits. The surface acoustic wave filter features a very compact unit. This device is accurate in establishing correct IF response. The length of the surface wave differs somewhat according to the material element of the piezo-electric element. The interdigital transducer is so arranged in length and width to provide the desired bandwidth for the PIX IF circuits. This small device very seldom causes any problems, and it takes the place of several transistor or tube IF stages.

Video detector and amplifier

Within the tube circuits, a duo-diode tube with plate, cathode, and heater elements formed the video detector. Then came the semiconductor diode for video detection. Today, video detection is accomplished inside the IF/video circuits before connection to the video circuits. Several stages of video are located in the same IC. One or two video transistors might follow the video circuits in today's TV chassis.

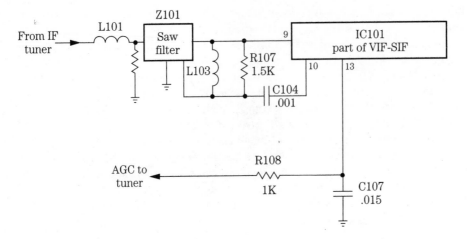

2-28 A block diagram of a saw filter network between the IF input and IF amplifier IC.

Solid-state video section

The transistorized video section might be included in one or two transistors, single IC, or included in one large IC with IF amp, sound detector, sound amp, video detector, video amp, noise canceler, and AFT detection and AGC detection circuits (Fig. 2-29). In some recent color TV chassis, only two large ICs are in the whole chassis, except the sweep output stages.

The IF signal from the picture IF filter (saw) might have one preamp transistor, or in recent color receivers the saw filter couples directly into a large IC. Internally, the IF signal goes through an IF amp or buffer stage, video detector, and then to the video amp. The output of the video amp is fed to a coil and 4.5-MHz trap ceramic filter, where the sound carrier component (4.5 MHz) is removed and applied to the video and color processor IC. The sound signal is capacity coupled before the ceramic trap filter and fed to the sound circuits.

Solid-state audio-detector-amp circuits

All sound circuits, except the audio output, are located in one large IF video IC. In some earlier audio circuits, a separate IC could process the IF sound and include a driver sound amp (Fig. 2-30). The FM sound detection is tuned with a small coil that can be touched up with the adjustment tool. When the sound becomes erratic or muffled, try adjusting the sound detection coil for clarity. Sometimes a change in temperature will cause the sound coil to drift off frequency.

Audio circuits

Beginning at the discriminator output and all the way to the speaker, the signal is known as *audio* or *sound*, because it consists only of audio frequencies. The early solid-state circuits utilized transistors. Today, you will find mostly IC audio circuits, although a few TV chassis still have transistors in the final audio output circuits.

A poorly adjusted discriminator coil might result in distorted sound. Sometimes the discriminator coil might drift with excessive moisture and require alignment with a insulated screwdriver. Simply insert the required alignment tool inside the core area. Do not use a metal screwdriver and tear up the ceramic iron core. Slightly turn the core one way or the other until the sound is clear and crisp.

Solid-state audio output circuits

Transistorized audio output circuits might be included with a preamp or driver stage within the large video IC or all included within the audio output IC. Some audio output amps have several audio transistors (Fig. 2-31). The audio signal from the

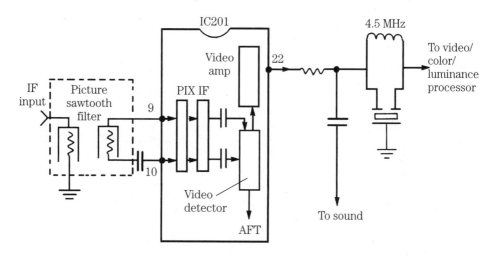

2-29 A block diagram of a typical IF section.

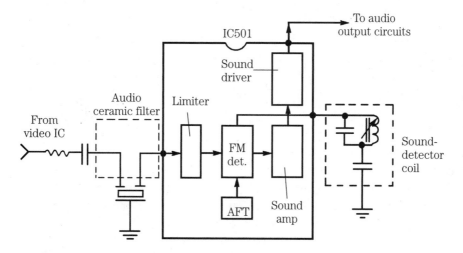

2-30 A block diagram of an IF sound section.

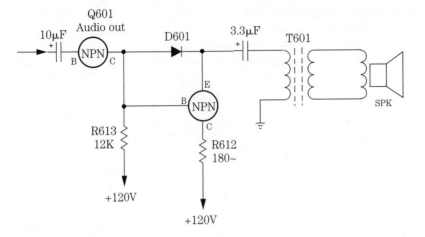

2-31 A block diagram of a typical transistor sound output circuit.

driver inside the IF sound IC is fed to another AF transistor that drives two transistors in push-pull operation.

Usually, when audio output transistors are used, a matching output transformer is coupled to the speaker. When the IC is used as audio preamp, driver, and output circuits, the speaker is coupled with a fairly large electrolytic capacitor. Check this coupling capacitor when the symptom is intermittent or dead sound. Stereo decoding in included in the high-end models.

The audio output transistors in the single-ended push-pull circuit are: one npn and one pnp. If the voltage at the collector of Q300 is decreased to below the specified limit, Q301 is turned off while Q303 is turned on. Thus, the current is limited at the emitter of Q303 and is connected through the transformer primary winding. Then, if the voltage at the collector of Q300 is increased to a limit, Q301 is turned on while Q303 is off, producing current through the primary winding of output transformer.

IC sound output circuits

The sound from the IF IC is coupled to the audio output IC201 through a small electrolytic capacitor. IC201 includes the audio AF and all audio circuits in one chip (Fig. 2-32). Often, a fairly high dc supply voltage is fed to the IC output circuits. The sound output is capacitively coupled to the speaker. Because one side of the speaker voice coil is grounded, the electrolytic capacitor isolates the dc IF pin voltage and couples the audio to the speaker.

Sync circuits

The purpose of the sync section or circuits is to provide proper timing and placement of the image bits on the TV screen. This synchronizing information is transmitted by the TV station as part of the picture signal. The two main types of pulses are those responsible for the starting time of each line on the main left side of the screen (horizontal sync pulses) and those controlling the exact start of the picture at the upper left corner of the picture tube (vertical sync pulses).

Thus, the first function of the sync circuits is that of a sync amplifier; that is, to increase the sync signals to their required level. The older TV sets have a sync separator stage that separates the vertical sync pulses from the horizontal pulses. There is, additionally, a third function called *sync clipping*, which, in simplified terms, means the extension of sync information from the combined picture-sync signal. Today, one IC does the work for the sync and AGC circuits.

The sync circuits have no effect on the sound, and only indirectly on the picture. Actually, the picture information is completely independent of sync performance; however, because the sync signal controls the sequence and location of each picture element, any malfunctions in the sync section are certain to produce chaos in the picture, either an up or down rolling of picture or tearing in a horizontal direction. Check the sync circuits when the picture will not stand still or stop sliding sideways.

Solid-state sync and AGC circuits

Within the tube chassis, a separate tube was used for sync and AGC circuits; in the transistorized chassis, separate transistors were used for each circuit. Today, the AGC and sync circuits are all contained inside the PIX IF and video IC. Internally, the RF and IF AGC circuits are taken from the IF amp and buffer circuits. The RF AGC is fed from terminal B to the varactor tuner. The AGC delay control is fed from pin 12 of the RF AGC circuit. This RF AGC delay control adjusts the input voltage to the tuner circuits (Fig. 2-33).

The latest AGC circuits are located in the IF/SIF/AGC/SYNC (IC101). The tuner AGC voltage is taken from pin 20, filtered, and fed to the AGC terminal of the tuner. RF AGC control (R111) sets the AGC voltage range at pin 25 of IC101. Notice the test point (TP12) to measure the IF AGC voltage (Fig. 2-34). The AGC circuits are all controlled inside IC101.

Vertical circuits

We know that the picture on the TV screen is "painted" by an electron beam moving in zig-zag fashion, one line at a time, from left to right. It is apparent from this action that as the beam sweeps from left to right, it is also pulled downward at a constant rate. This downward movement of the beam is accomplished by a system

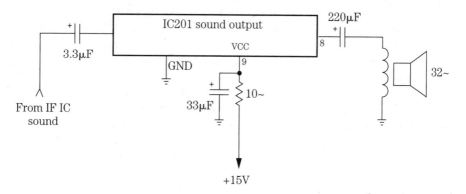

2-32 The audio output circuits with a sound output IC.

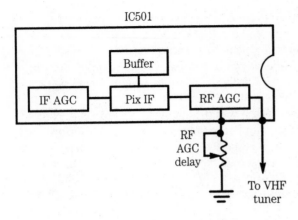

2-33 A block diagram of typical AGC circuits within the IC.

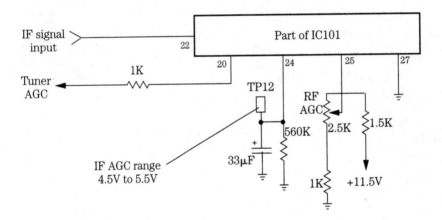

2-34 The AGC circuits are contained in the IF/SIF/video IC.

known as the *vertical sweep*. The vertical section consists of an oscillator and amplifier. The signal is then fed to the vertical output transistor or IC.

No vertical sweep causes a single bright white line across the center of the screen (Fig. 2-35). Insufficient vertical sweep might result in a four- or five-inch picture. A bright line at the top of the picture, with bunching horizontal lines, indicates problems within the vertical circuits. A crowded (compressed or squashed) picture at the top and a stretched picture at the bottom are often caused by a defect (nonlinearity) in the vertical output circuits. Vertical foldover is usually caused by a pincushion or feedback circuit in the vertical output circuits.

Three adjustments are associated with the vertical sweep system, which might require resetting in case of a malfunction. These are vertical hold, vertical size, and vertical linearity. The first of these in most sets is accessible from the front of the cabinet. The last two almost always are located in the rear because they seldom require adjustment. Each is described under troubleshooting, where cause-and-effect relationships are outlined.

As the name implies, the vertical hold is used to prevent rolling vertically. The vertical size is an adjustment to make the picture cover the full height of the visible screen. The vertical linearity adjustment serves to adjust the picture for minimum distortion, as when the lower half of a circular object looks flattened while the upper half is egg-shaped. The vertical linearity control is adjusted to make a test pattern circle perfectly round.

Solid-state vertical circuits

Separate transistors can be used in the early transistorized chassis within the vertical circuits. A sawtooth signal is fed from the luminance/color IC and applied to the base terminal of the vertical drive transistor. This causes the vertical output transformer to conduct. Both vertical output transistors are conducted by the sawtooth signal produced at the vertical drive transistor. The output is capacitor coupled through the electrolytic capacitor to the vertical deflection yoke winding (Fig. 2-36).

You might find the vertical sync, ramp generator, and vertical driver in one large luminance/color IC. The ramp generator signal is taken from the second countdown circuit. The vertical driver signal is taken from pin 27 and applied to the input vertical output IC. The vertical output pulse from IC501 is connected directly to the vertical yoke winding (Fig. 2-37). In some TVs, the vertical output is included within the deflection IC; in others, two vertical output transistors are tied to the vertical driver circuit within the deflection IC.

Vertical Output Circuits

Today, the vertical output circuits consists of one large IC. The higher supply voltage is fed to the vertical output IC than transistor circuits. Most vertical output ICs are mounted on a heatsink and can run quite warm compared to other ICs. A red-hot vertical IC must be replaced.

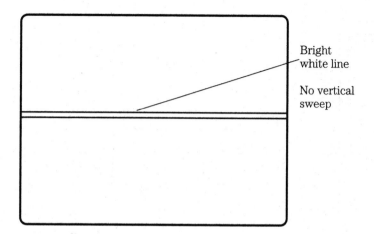

Bright white line

No vertical sweep

2-35 A white horizontal line indicates that there is no vertical sweep.

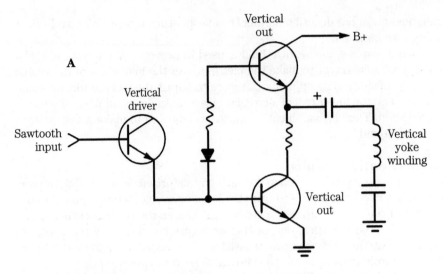

2-36 A block diagram of a transistor output vertical circuit.

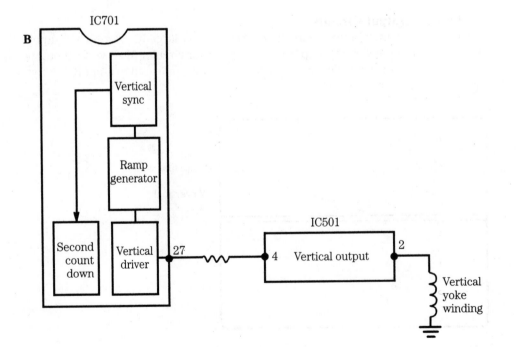

2-37 A block diagram of an IC countdown vertical circuit.

The input signal from the deflection IC is fed to pin 6 of the vertical output IC402. Pin 9 is the voltage supply terminal (Fig. 2-38). The output from pin 2 feeds directly to the vertical yoke windings and a 1000-μF electrolytic capacitor with a low-value resistor is contained in the return leg of the vertical deflection yoke. This vertical seep signal is amplified many times to drive the vertical yoke winding. Check the waveform into the output IC and at the output terminal. Check all connecting components for defects, before replacing the vertical IC.

Horizontal circuits

The horizontal circuits consist of a horizontal oscillator, and a countdown or deflection IC circuit. The deflection IC provides both vertical and horizontal drive pulses to the respective vertical and horizontal stages. The horizontal drive transistor can amplify the horizontal drive signal to 7 to 10 times the input signal. This drive signal or waveform is transformer coupled to the horizontal output transistor (Fig. 2-39).

The collector terminal of the horizontal output transistor is contained in the primary winding of the flyback or horizontal output transformer. The voltage applied to the collector terminal is supplied through the primary winding. The flyback supplies high voltage to the picture tube, screen, and focus voltage. In the latest TV chassis, the IHVT (integrated) transformer provides several voltage sources to the various sections of the TV.

Solid-state horizontal circuits

The solid-state horizontal circuits consist of a horizontal oscillator, driver, and output stage. In countdown deflection, the horizontal oscillator feeds into the first countdown circuit and feeds a horizontal predriver stage (Fig. 2-40). The horizontal sawtooth waveform is fed from the predriver (pin 17) to the driver transistor (Q401). The horizontal pulse is transformer coupled to the horizontal output transistor (Q402).

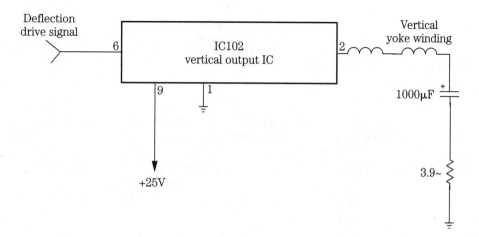

2-38 A typical vertical IC output circuit with a yoke-return capacitor and resistor.

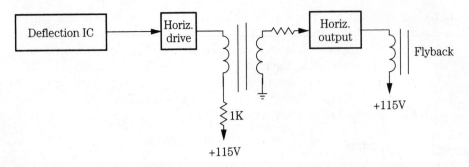

2-39 A block diagram of a horizontal output circuit with a flyback transformer.

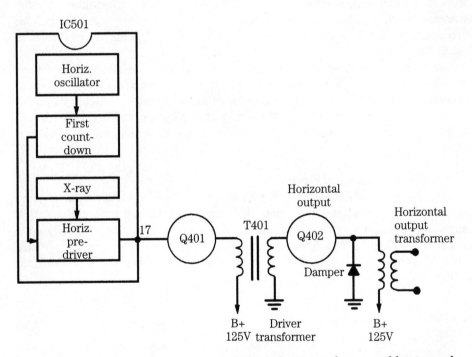

2-40 A block diagram with a horizontal countdown IC, driver, and horizontal output transistor. .

The horizontal output transistor generates horizontal scan and high voltage to the CRT. Actually, the horizontal output transistor acts as a switch during horizontal scanning. The output of the horizontal output transistor applies a high peak pulse to the damper diode and horizontal deflection coil. The flyback or horizontal output transformer is connected to the output of the horizontal transistor (Fig. 2-41).

Most problems within the solid-state TV chassis are caused by the horizontal output transistor. The transistor will either short, or appear leaky or open. The flyback transformer and drive pulse should be checked when the output transistor appears leaky or shorted.

High-voltage circuits

The high-voltage (HV) of any TV set is physically distinct and separated from the open part of the chassis, primarily for reasons of safety. Here the extremely high potential (30,000+ volts, in many color TV sets) for the picture tube is developed. Although it is true that the energy behind this extra high voltage is not enough to be lethal, the danger from this source is nonetheless very serious. Even secondary effects, such as the reaction to the shock, a resultant fall, etc., can be dangerous. Caution labels in the vicinity of the high-voltage cage have their purpose, but a clear understanding of what is involved is even more valuable.

At this point, there is a need to elaborate somewhat on the mechanism of "picture painting," to make some troubleshooting procedures and corrective adjustments more understandable. The picture tube of any set, monochrome or color, consists of three functionally distinct components: the electron gun, the positioning of deflection structure, and the screen.

The gun is physically located in the rear or neck portion of the tube, in which the glowing heater can be seen when the set is on. It performs the function of generating, shaping, and focusing the electron beam to a pencil-point sharpness at the point where it hits the screen. The image-brightness adjustment is also connected to this portion of the tube. This is an invisible beam of electrons, not light. The light you see

2-41 The location of the horizontal output transformer with focus and screen controls.

on the screen is the result of the electron beam striking a phosphor coating inside the picture tube.

Positioning of the beam, including its zig-zag movement across and down the screen, is accomplished by the outputs of the horizontal and vertical amplifiers connected, respectively, to the vertical deflection yoke (coils placed around the picture tube neck). It is, therefore, obvious that if the beam stays on a single horizontal line, instead of moving gradually down to create a complete picture, the vertical deflection system is at fault. By logical deduction, you might conclude that a failure in the horizontal deflection system would produce a single up-and-down line on the screen. Logical as this might seem, this is not the case because of that second important function of the horizontal amplifier referred to earlier.

The third portion of the picture tube, the screen, is an electron-to-light converter. The coating on the inside of the tube has this capability, but it will produce light only if the electron beam strikes the coating at sufficiently high speed. This high speed is imparted to the beam by a high voltage (actually many thousands of volts) generated by the horizontal amplifier under normal operation; therefore, whenever the horizontal amplifier is not performing, no such high voltage is generated and no light whatsoever appears on the screen.

Another key device in some sets is a fuse, not for the TV set as a whole, but only for the horizontal deflection system. Failure of this fuse, sometimes without apparent cause, will disable the horizontal amplifier, (indirectly) the B+ boost, and the high-voltage power supply. The symptoms will then be the same as a failure of any other vital link in block L, that is, the loss of all light on the TV screen. Failure of this fuse is rather infrequent, and because it is a part of the rather dangerous high-voltage system, it is not accessible from the outside rear of the cabinet, as is the main 115-volt ac fuse. In addition, it is often soldered in place. Fortunately, the replacement can be made in almost all TV makes without the use of a soldering tool.

A few operating adjustments can be performed in the horizontal deflection system. These adjustments are reasonably accessible, quite safe, and (with use of the manufacturer's instructions where available) can be easily made. The adjustments are made at the factory and might require minor touching up after long periods (two years or more) of operation; or they might need to be readjusted after the replacement of a defective part.

Transistorized HV circuits

In the early transistorized TV chassis, the flyback transformer was quite small, with a tripler unit used to build up the high voltage. Today, the horizontal output transformer provides picture scanning through the horizontal yoke, high-voltage, and several derived-voltage sources. The high voltage is developed with integrated diodes molded right inside the transformer (HVT).

The flyback transformer is connected to the horizontal output transistor and B+ power source. The secondary winding produces very high voltage with the enclosed diodes between windings. Besides furnishing HV, this winding is tapped to produce the screen and focus voltage for the picture tube (Fig. 2-42).

Besides the HV winding, additional secondary windings can provide a high voltage (+250 V) for the picture tube and color amps. Sometimes another winding pro-

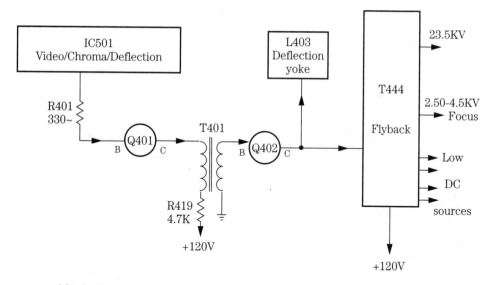

2-42 A block diagram of a TV with horizontal and HV circuits.

vides rectified voltage sources to the sound, vertical, luminance/color processor, and IF/video stages. Each secondary winding includes a small silicon diode with a filter capacitor for these secondary derived voltage sources. You might find the heater voltage for the picture tube loosely wound around the ceramic frame of the flyback transformer.

Today's high-voltage circuits

The present-day TV chassis contains a integrated high-voltage transformer (IHVT) with several secondary voltages. High-voltage diodes and capacitors molded inside the flyback form are used to develop high voltage. This high voltage is applied to the anode terminal of the picture tube or CRT. The larger the physical size of the screen, the higher the voltage. Besides high voltage, the horizontal output transformer or flyback supplies focus and screen voltage to the CRT, through a divided voltage network (Fig. 2-43).

Several secondary voltages can be added to the flyback for scan-derived voltage sources. These secondary voltages can have IC, transistor, or zener voltage regulation, or a combination of a transistor and zener diode circuit. Each winding can have a separate silicon rectifier or diode, and electrolytic filter capacitors. The secondary voltages can provide power to the audio, vertical, deflection IC, video, IF, chroma, luminance, and picture tube circuits (Fig. 2-44).

Internal damper diode

The damper diode is used in the horizontal output transistor collector circuits. A damper diode prevents oscillations in the horizontal output circuits and ringing in the low-voltage power supply. Today, the damper diode is found inside the horizontal output transistor. Be careful when taking resistance or diode-junction

2-43 The integrated high voltage transformer (IHVT) contains HV capacitors and diodes in the molded form.

DMM tests on the new output transistor with a damper diode connected internally (Fig. 2-45). A normal measurement in one direction might be mistaken for a direct short from collector to the emitter terminal, with a damper diode inside.

Always check the damper diode when you find a leaky output transistor. An open damper diode can cause the transistor to overheat and be destroyed. A high leakage between collector and emitter terminals can indicate a leaky damper diode or transistor, if a low-resistance measurement is found in both directions. Likewise, a resistance measurement between chassis and the metal case of the output transistor can indicate a leaky damper diode or horizontal output transistor. A leaky damper diode can be checked with the diode test of a DMM.

Horizontal width

In older TVs, horizontal width was adjusted with a variable resistance or width coil. You might still find a width adjustment in today's TV portables. Yet, you can find horizontal pincushion transformers or coil adjustments that can change the width in large TV screens. Improper width can result from low voltage applied to the horizontal output transistor, defective hold down or safety capacitors, or improper adjustments of the pin-cushion circuits. A white vertical line (up and down) indicates that the sweep or yoke winding is open (Fig. 2-46).

Horizontal hold

This adjustment is physically very much like the width adjustment. Its purpose is to preset the circuit controlling the movement of the beam across the screen (left to right) to its approximately correct position. Then, the automatic frequency control (AFC) can take over and make the timing precise. In some cases of picture tearing, a slight adjustment of this circuit will restore the picture to stability. But the horizontal hold control cannot correct any sweep faults. Further, when an adjustment is made, it must be done very gradually, sometimes only a fraction of a full turn, for results.

Horizontal drive

Some older sets contain an adjustment marker H drive (for horizontal drive). Hopefully, most of these sets have passed on to the big trash dump. This adjustment has the effect of stretching one side of the picture and should only be made when poor horizontal sync exists. Don't try this adjustment unless you really feel that it will help. Most TV servicers only attempt this with proper test equipment; however,

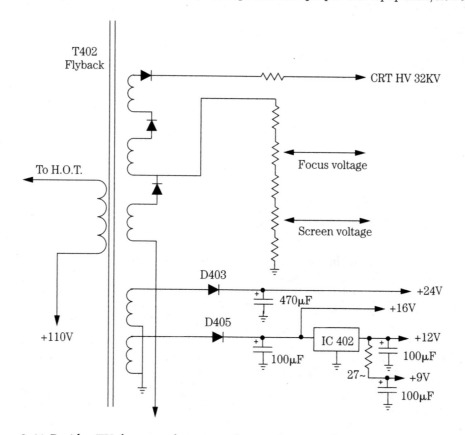

2-44 Besides TV, focus, and screen voltages, the secondary voltages of the flyback can power other TV circuits.

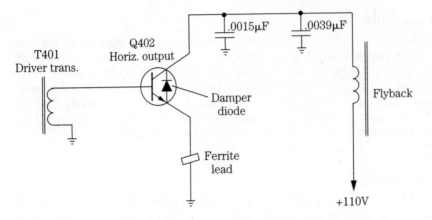

2-45 Today, the damper diode is mounted inside the horizontal output transistor.

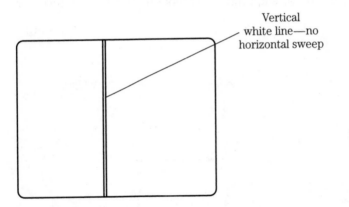

2-46 A white vertical line indicates that a yoke winding is open or that there is no horizontal sweep.

there is an alternate method of adjustment that required only a screwdriver and good eyesight.

Tune in a very weak station, one with lots of snow. Turning the adjustment in one direction will produce a white vertical (up and down) line in the center on the screen. Now turn the adjustment until the line just disappears, and you've got it.

Horizontal lock-in

This adjustment, like the previous one, is found only on some sets. Its behavior is somewhat similar to the hold control described earlier. Turning it from its normal position toward either extreme will produce a diagonal tearing of the picture, in one direction or the other, depending on the direction of rotation of the adjustment. Figure 2-47 is an example of such tearing. The lock-in control's correct setting is the point where channel changing or adjustments will not produce any tearing because of the automatic control by the horizontal sync pulse.

Centering controls

These functions used to be rear-of-chassis adjustments on TV sets a number of years ago. They are no longer in use today because they were replaced by a mechanical centering device, which is part of the picture tube deflection yoke. However, in those rare cases where one of these adjustments is located on the back of the chassis, the procedure is simple. With a test pattern on the screen (or an equivalent circular emblem of some sort), the control is adjusted for best centering of the image on the screen.

Low-voltage power supply

Today's low-voltage power-supply circuits operate directly from the ac power line. A 3- to 5-A fuse protects the power line and components in the low-voltage circuits. If a part becomes leaky in the bridge rectifier, filter capacitor, IC regulator, transistor regulator, and horizontal output circuits, the main fuse can blow or open. When the on/off switch is on with a manual remote switch, ac voltage is applied across the thermistor and degaussing coil. The picture tube is degaussed to erase impurities upon the TV screen each time that the TV is turned on (Fig. 2-48).

The bridge rectifier circuits can consist of four separate silicon diodes or one bridge component. C105, a high-rated capacitor with high working voltage (220 μF) provides raw dc filtering. Notice that the dc voltage at this point has increased to almost 160 volts, compared to the 120-volt ac power line voltage. IC101 is a high-voltage regulator IC with a fixed output at 115 volts. Here, a dc fuse or fuseable resistor can be installed to protect the dc circuits if the horizontal output transistor, transformer, or flyback becomes leaky or shorted. Several different voltage sources can feed to the horizontal and vertical deflection circuits from this same low-voltage supply.

Solid-state power supply

Many different power supplies (half-wave, full-wave, regulated, pulse-width-modulated, and chopper circuits) are contained in the solid-state TV chassis. In a typical half-wave or bridge rectifier, the low-voltage power supply operates directly from the power line, but the power transformer provided ac voltage in the tube receiver. The ac line is fused with another fuse after the half-wave rectifier. A large filter capacitor helps filter out the ac from the dc source.

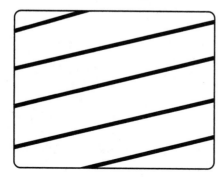

2-47
Horizontal slanted lines
indicate that the horizontal
oscillator is off frequency.

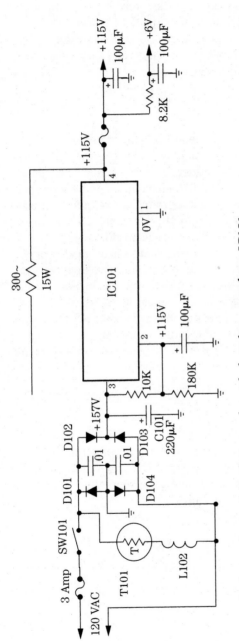

2-48 The low-voltage power-line supply with dc voltage regulator IC101.

The rectified dc voltage (145 V) is fed to a high-voltage regulator IC. This regulator is built somewhat like the horizontal output transistor or power sound output IC. Sometimes these regulator ICs are shunted with a high-wattage resistor to the output-regulated terminal (115 V). This dc source is fed directly to the horizontal output transistor, sound output, and deflection IC circuits (Fig. 2-49).

The pulse-width chopper power supply can consist of a full-wave bridge rectifier, regulator drive transformer, regulator control IC, x-ray protect, overcurrent shutdown transformer, chopper output transistor, and chopper output transformer. The output voltage from transformer provides a 125-, +24-, -24-, and a 16-volt source. The 125-volt source powers the horizontal output circuits.

Hybrid TV set

When hybrid chassis were designed, transistors were not capable of operating at the high-voltage potential required by some circuits. Therefore, transistors were used in the low-voltage circuits (tuner, sync, AGC, IF, color, and video driver stages) of the hybrid chassis. Transistor circuits require less power than the circuits of their tube counterparts.

The tubes in the hybrid chassis have filament voltages rated at 25, 38, 40, and 42 volts ac. If the filament voltages of the tubes in the chassis are added together, their sum will often equal the line voltage of 120 volts ac. In hybrid sets, these tubes are used in the horizontal and vertical sweep circuits. Some hybrid chassis also use tubes in the video output and color demodulator circuits. Because a majority of TV problems occur in the sweep circuits, many troubles can be resolved by replacing a vacuum tube.

Heat and light play an important role in diagnosing faulty tube circuits. Because transistors do not use filaments or heaters, another method of locating suspected stages is required. Transistor stages are often fabricated into small subassemblies containing a number of associated parts mounted on a printed-circuit board (PC board).

Degaussing circuits

The degaussing circuits counteracts the earth's magnetic field and any magnetic or electromagnetic force that can magnetize the TV screen. Large audio speakers near the TV can magnetize the front screen. When sweeping the carpet in front of the TV set, if the sweeper is shut off, this can induce magnetic impurities into the TV screen. The degaussing circuits remove these impurities each time that the TV is turned on.

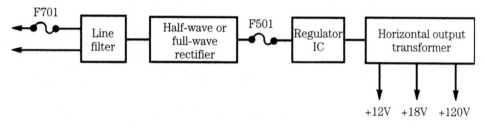

2-49 A block diagram of a typical low-voltage power supply.

When the ac switch is turned on in the TV set, ac voltage is applied to the cold thermistor and degaussing coil, as both are in series and connected across the 120-volt power line (Fig. 2-50). Usually, when the thermistor is cold, the resistance is from 5 to 20 Ω, letting the ac voltage applied across the degaussing coil as it provides a magnetic force to remove impurities from the TV screen. As the thermistor becomes warmer, the resistance increases until no voltage is applied across the degaussing coil. When the TV is turned off, the thermistor restores to its original low resistance.

The degaussing coil is mounted around the front edge of the picture tube. Often, the degaussing coil can be unplugged when the chassis is removed for service. In some cases, you might be required to use a large hand-held degaussing coil to remove very strong magnetic forces from the TV screen.

Integrated circuit applications

The integrated circuit (IC) is replacing the discrete component circuit in many fields of electronics including television. Briefly, an integrated circuit is an assembly of various parts, including transistors and diodes, so that only the necessary connections are externally accessible; all other interconnections are internal and permanent. The physical size of the IC package depends on the complexity of the circuit. The smaller devices have fewer circuit functions and need just a few external connections (pins) called *SSI (small-scale integration)*. More complex circuits capable of performing a multitude of tasks require a greater number of external connections (pins), called *MSI (medium-scale integration)* and *LSI (large-scale integration)* ICs.

In today's TV receiver, it is not uncommon to find just one IC replacing several individual circuits. One IC could replace all of the sound circuit or just the pre-amp and driver stage. Individual ICs could replace the AGC, sync, and video IF circuits, or you could find all of these functions replaced by one IC. The horizontal and vertical drive circuits might require separate ICs in one receiver and have both functions performed by one IC in another receiver. Depending on the make and model, ICs are replacing discrete circuits in many of today's video and color circuits, and in the tuner, memory, and tuner control circuits, as well (Fig. 2-51).

Replacement of a single IC could solve many different receiver troubles, but even the failure of one section of the IC would require the entire IC to be replaced.

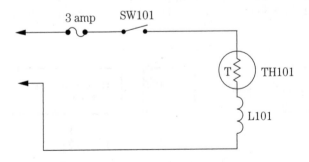

2-50 The degaussing coil is energized each time that the TV is turned on.

Large
IC

2-51 One large IC contains most of the TV's circuits.

Although this might at first seem uneconomical, actually this is not the case. Pre-assembly and encapsulation save more in the long run, making it more economical (and often more reliable).

In Chapter 5, we recommended that the TV owner purchase the complete package of data for his or her particular model receiver. In such a package, there always are schematic diagrams and other partial diagrams necessary for servicing the receiver.

Printed-circuit wiring

Today, the main chassis is located on PC wiring or an etched wiring board. In fact, the first TV manufacturer to incorporate PC wiring was Admiral Corporation. The printed circuit is placed on a very thin copper sheet cemented to a fiberboard. After the circuit is laid out, the copper-clad board is etched.

The liquid etching solution is either ferric chloride liquid or ammonium persulphate powder etching powder. The board is etched after passing through this solution in the automatic machine. The board components are placed into the correct holes and passed through a bath of solder. All connections on the board are soldered and processed with one machine.

The PC board can be plugged into sockets or connecting wires attached for other larger components (Fig. 2-52). Some of these boards are double-sided with regulator components on top, and surface-mounted parts underneath. If the PC board is too thin, it might warm and pop some rivets or PC wiring connections. Most of the trouble with surface-mounted boards are poor connections between parts and PC wiring.

The PC board might crack or split if the TV is dropped or when handled improperly. Look closely around heavy components on the board for cracked areas. A poorly soldered connection might be located at the bottom of component wire or ground eyelets. When replacing large components on the board, be careful not to damage PC wiring. Too much heat applied to the PC wiring might cause it to lift up from the board. Clean off all rosin residue from the connections with circuit coolant, contact cleaner, or rosin flux remover.

Surface-mounted components

Surface-mounted components have been used in compact disc players and camcorders for several years and have just entered use with the TV chassis. The IC processors and control ICs are available in flat-pack IC packages with terminal leads coming out of the sides of the component. These leads are soldered directly to the PC wiring.

The IC processor might contain as many as 82 terminal leads in a camcorder, but the transistor surface-mounted component has only three terminals. Of course, the resistor and capacitor have only two flat-end terminals. The digital transistor has only three leads with base and bias resistance enclosed inside. You might find a dual

2-52 The printed wiring forms the complete TV chassis in a recent TV.

2-53 Surface-mounted components are used on the PC wiring at the bottom side of the chassis.

resistor network with two separate resistors in one chip with solid soldered connections at the ends. These leadless chips must be heated at the end with the soldering iron and removed with a pair of tweezers.

With a double-sided PC board, the larger components for the TV circuits are located on top of the chassis and surface-mounted components underneath. The fixed resistors and capacitors are quite small in size. Figure 2-53 shows a card of fixed resistors with five of one value. When a double-sided board is used with surface-mounted components underneath, the overall TV chassis is quite small.

3
CHAPTER

Color TV and how
it works

Up to this point, color TV has not been specifically covered, but everything that has been said pertains equally to color. In fact, when I mention the circuits and functions peculiar to color receivers only, you will see that these circuits and functions are additional to the basic TV set circuits. It is necessary to adjust one control (color) to its minimum (equivalent to off) position in any color TV set to arrive at a black-and-white representation insofar as results are concerned.

Generally speaking, all the TV circuits up to and including the video detector-amplifier are common to all signals—sound, video, and color.

The visual spectrum

Figure 3-1 is a graphic depiction of the color components of visible (white) light after passing through an optical device called a *prism*. Although discrete color blocks are indicated, all colors gradually blend into each other. Figure 3-2 is a much simplified version of the same phenomenon, more familiar to the student of elementary painting. The three primary colors produce intermediate colors as well as white. In the present color TV system, this basic three-color system is utilized to produce the picture.

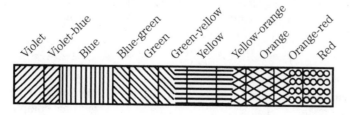

3-1 Pure white light is composed of all the colors depicted here.

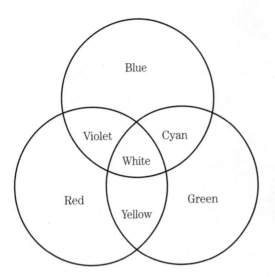

3-2
The color wheel principle shows how the three primary colors blend into these intermediate colors and white.

Color transmitter

Figure 3-3 is a functional or block diagram illustrating (in much simplified form) the components of a color picture transmission system. A comparison with the monochrome TV transmitter in Fig. 2-6 shows the basic difference between the two. Consider the components of a color system.

Camera tubes

Three camera tubes (block A, Fig. 3-3), one for each of the primary colors together with their individual amplifiers, combine to produce the three components of color information required to produce a color picture on the home TV set. These three components are:

- *Picture brightness information* This establishes the overall color picture brightness in the TV set, not just the brightness of any particular color, but also the background brightness level (proportionately) correct for all colors.
- *Picture color information* This is the heart of the color signal and contains all the information required by a TV set to reproduce the entire range of colors, corresponding to the original scene.
- *Color sync information* This corresponds (and is in addition) to the earlier-described sync functions in black-and-white television. It is sometimes referred to as the *3.58-MHz color burst*.

Modulators

The modulator or combiner portion of the color TV transmitter corresponds to the similar functions in the monochrome transmitter. Here, the various electrical components of the color picture are combined (block B, Fig. 3-3), without losing their identities, into an overall color signal, which, in turn, is combined with the sound signal.

Final amplifiers and antenna

The final amplifier and antenna system are essentially the same as the black-and-white transmitter in Fig. 2-6, except for the addition of the color information. Neither the final amplifier nor the transmitting antenna "know" the difference; that is, they amplify and radiate (respectively) a total television signal that consists of the required components. This same is true with a receiver.

Color receiver

The color TV receiver is a combination of a basic monochrome receiver plus an add-on section that only deals with the color functions of the set. In fact, this add-on section picks up where the black-and-white portion of the set leaves off. Consequently, and as was briefly stated earlier, the standard procedures for analyzing and troubleshooting a color TV set is to first consider it as a black-and-white set. This is easily done by reducing the color signal (turning the chroma control) to zero, thus leaving a black-and-white picture, as would be received by a monochrome receiver tuned in to a color program. It is for this reason also that the block diagram to follow (Fig. 3-4) is divided into two sections by a dashed line. The black-and-white portion is readily recognizable as the familiar monochrome TV set shown in Fig. 2-6; therefore, the black-and-white section receives only a minimum of emphasis in this section.

It is advantageous, however, to view the complete TV receiver as a unit, to clearly visualize the continuity between and the transition from the combined monochrome and color functions to the strictly color portions and functions. The overall functional diagram is represented with this purpose in mind.

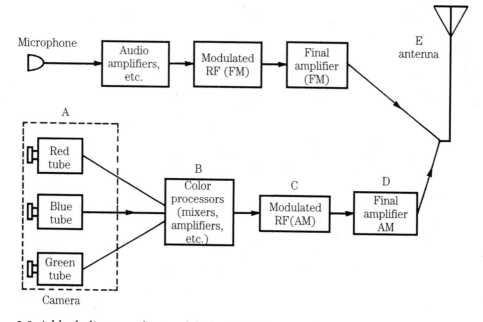

3-3 A block diagram of a simplified color TV transmission system.

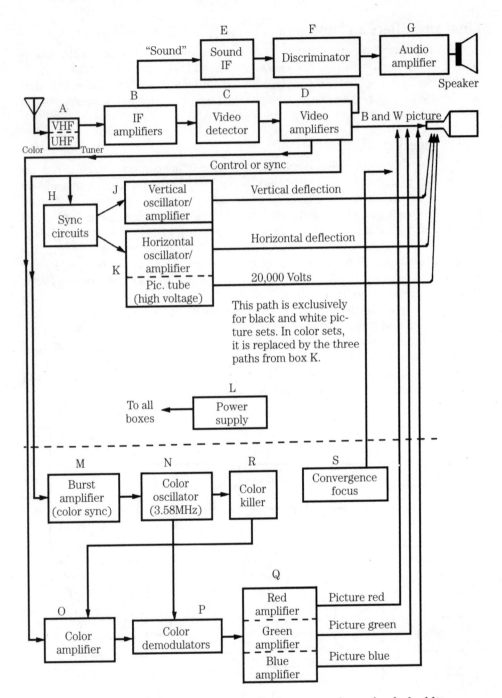

3-4 A block diagram of a color receiver. The functions above the dashed lines parallel those in a black-and-white receiver. Below the dashed line are color-only functions.

Figure 3-4 gives a simple, nontechnical means of comparing a monochrome TV and a corresponding color set. It shows a complete color TV, with no reference to black-and-white operation; however, it is obvious that the complete black-and-white TV of Fig. 2-6 appears with hardly any change inside the framework of the color set diagrams.

In addition to the basic facts of structure, this graphic comparison also shows, in greatly simplified form, the two main constituents of the color picture. Function blocks O and P are responsible for the three component colors, plus the overall color picture brightness information. Function blocks M and N provide the information for the proper color registration and mixing of the three primary colors into the complete range of color shading, which makes up the final color image on the home TV screen.

Before covering the individual stages or functional blocks of the complete color TV set, briefly analyze the complete sequence of functions in Fig. 3-4 to get a concise look at what happens. The TV signal, containing picture information and sound information, is intercepted by the antenna. It is a composite (all-in-one) signal. Block A (the tuner) selects any one desired station from the number of stations available in a particular location. The IF (intermediate frequency) amplifiers (block B) amplify the selected station to the signal level required for further processing.

So far, the composite signal picked up by the antenna and selected by the tuner has been made larger as required, but it still is riding piggyback on a carrier. Block C, the video detector, performs one simple main function: it separates the intelligence (picture, sound, and sync information) from the carrier, which served as a vehicle for transmission of the signals. The output of the video detector is the composite signal described previously.

Although the carrier appears to be discarded at this point, it is not. As far as the intelligence goes, it has done its job. Nevertheless, it is utilized to perform the vital function of adjusting the TV receiver's amplification to the required level; that is, the weaker the incoming signal, the more it is amplified, and vice versa. This is known as *automatic gain control (AGC)*. Incidentally, the AGC is a back-of-the-set adjustment that sometimes requires resetting on both color and black- and-white receivers.

Block D in Fig. 3-4 is the video amplifier. True to its name, it amplifies the video signal together with its other components (sound and sync signals) received from the video detector just described. At the output of the video amplifier, the three signal components separate, each going its own way.

Picture information (commonly called *video*) goes to the picture tube for conversion into an optical image. Sound information, a sound-modulated carrier (sound energy piggybacked on an RF carrier), goes to block E—the sound IF amplifier. The signal now is the equivalent of an FM radio signal in an ordinary FM receiver. As such, it is a sound signal superimposed on a carrier. This in no way contradicts the earlier statement that a carrier has been discarded by the video detector, although the technical explanation is of no consequence to our purpose. After amplification the signal goes to an FM detector (block F) or discriminator, where sound energy only is extracted. Block G is an audio amplifier, which brings the soundenergy to the level required by a speaker connected to the output of block G.

At this point, you might be wondering why the same amplifiers appear over and over again, interrupted by other functions in the chain. Wouldn't it be better to have just one amplifier in one place and do all the required amplification at once? It would seem simpler; however, a TV (and radio, too) receiver is designed to receive a faint signal from the antenna. In fact, such a signal might be one million (or more) times weaker than the voltage of the ordinary flashlight cell (nominally 1.5 volts). Necessary incidental losses within the TV set make this situation even more crucial. In other words, the incoming TV signal must be amplified millions of times before it becomes the final product (picture or sound).

The most practical and most (technically) efficient way to achieve this tremendous magnification is to do it in small stages and at particular locations in the set. Thus, some initial amplification (RF amplifier) is done immediately after the signal arrives from the antenna. Additional amplification is provided after initial processing (conversion to IF in the tuner), this time in the IF amplifiers. A third and usually final portion of amplification is accomplished, after another conversion (detection), in the video amplifier for the picture signal and in the audio amplifiers for the sound signal. There also are other amplifiers in the TV set, for such nonsignal functions as sync and AGC (automatic gain control).

Meanwhile, back at the video amplifier, the third component of the composite signal from the video amplifier is the *synchronizing-timing signal (sync)*. Block H contains sync amplifiers, separators, and shapers, with an overall function of making the timing pulses suitable for controlling the accuracy of the sweep voltages via the horizontal and vertical oscillators. Block J is the combined vertical/amplifier, which produces the voltage for vertical deflection of the picture tube beam. Block K is the corresponding group for horizontal deflection, as detailed earlier for monochrome. The dotted line in block K separates the strictly sweep function from the incidental (although most important) functions of B+ boost and high-voltage generation. Finally, block L represents the power supply for the entire TV set, including some adjustments and controls for the picture tube.

So far, we have described a complete black-and-white receiver, although hidden within its various stages were color signals—information convertible to a color picture after proper processing. The following paragraphs describe such conversion and processing.

From the video amplifier (block D, Fig. 3-4), the two components of the color signal go to the color-only portions of the receiver. Color picture information goes to block O, a color amplifier. After the signal is amplified to the required level, it goes to block P, the color detector or demodulator, then to the individual color amplifiers in block Q. Here are three separate amplifiers, one for each of the primary colors, whose outputs are applied, respectively, to each element of the three- color gun in the picture tube.

The second component of the color signal is the burst information. It is responsible for the proper color registration in the picture. It is fed to block M, the burst amplifier. From here, it is used to synchronize the color oscillator for precise timing or synchronizing of this all-important component of the color process. Although block M of the color section is analogous to block H (sync circuits) in the black-and-white portion of the set, the color oscillator is particular to the color process. No corresponding function exists in the black-and-white receiver.

The color oscillator signal (3.58 MHz) is used in the color demodulator (block P) to retrieve the three primary color signals. If each color is to be reproduced accurately, the color oscillator must operate in step with a similar oscillator at the transmitter. A failure in the color oscillator causes a complete loss of the color in a picture.

The color oscillator output is used as a carrier (just like a picture or sound carrier) to detect or demodulate the color signals. The color signal is called a *subcarrier* because it is within the black-and-white picture carrier.

The three additional functions in the strictly color portion of Fig. 3-4 are the color killer and the focusing and convergence systems (blocks R and S). Of necessity, these terms are explained further on.

Solid-state TV chassis

In the early solid-state TV chassis, only the tuner and sound circuits contained transistors. Some of these chassis were known as *hybrid TVs*. Then, the horizontal output section added power transistors. Afterward, the integrated circuits were introduced (Fig. 3-5). Again, the IF and AGC and sound circuits contained ICs. Next, the integrated circuits were used in color and luminance circuits. Today, you might find only two large ICs with very few outside transistors (Fig. 3-6).

The tuning section in recent color TV chassis might have a few transistors within the tuning, band selector, and AFT sections (Fig. 3-7). Some chassis might have the whole front end in ICs. The preamp transistor can be followed with a SAW filter and tied into a large IC that contains the IF amp, sound IF amp, sound detector and amp, video detector, video amp, noise canceller, AFT detector, and AGC detector.

In the output section of the solid-state chassis, three color transistors feed the CRT, with horizontal driver and horizontal output transistors (Fig. 3-8). The vertical output and horizontal deflection circuits can be a separate transistor or included in a large IC (Fig. 3-9).

3-5 Today's TV consists mostly of ICs, and very few transistors.

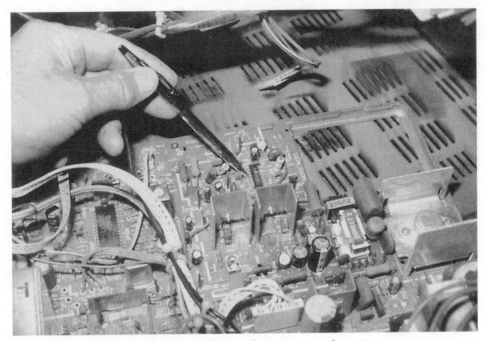

3-6 Here, a couple of ICs provide many different circuit functions.

One large IC processor might contain the video amp, pedestal clamp, chroma amp, color demodulator, AGC, sync separator, x-ray protector, horizontal oscillator, horizontal drive, vertical oscillator, and vertical amp. The vertical drive and output might be contained in one single IC. Today, most solid-state chassis have more ICs than transistors, such as the RCA CTC131 (Fig. 3-10).

Digital system control

In some TV chassis, the digital system-control circuits contain a microcomputer system that controls data in and out, clock timing, and system control reset features (Fig. 3-11). The system-control system microcomputer contains random-access memory (RAM) and read-only memory (ROM). The RAM section stores information for frequency channel change and customer settings, such as current picture and volume range. This allows the operator to turn on the last-viewed station with the same volume and picture adjustment when the receiver was turned off. The ROM section of memory stores information that is never changed.

The system control reset is to ensure that the digital microcomputer starts at the same time in the program each time power is applied. System-control communications controls the major functions within the chassis. The customer-control interface circuits control the function operating from system-control system. These controls consist of brightness, sharpness, contrast, color, tint, volume, balance, treble, bass, and power (Fig. 3-12). All these controls can be operated manually or on the hand-held remote control unit (Fig. 3-13). These digital control features are often included in higher-priced TVs.

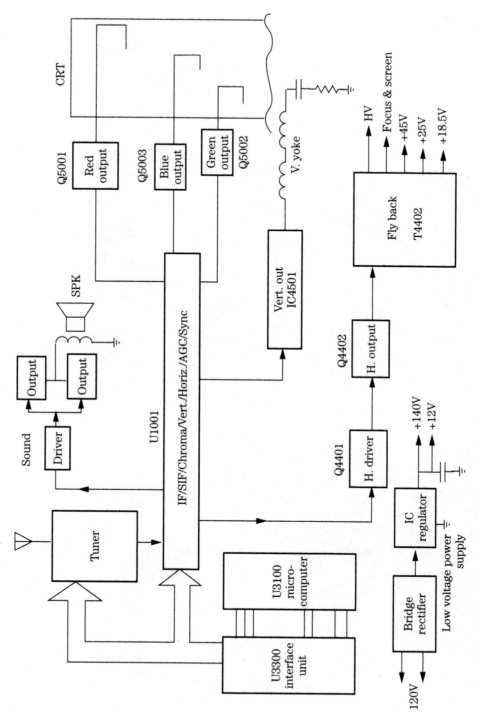

3-7 A block diagram of a RCA CTC146 TV.

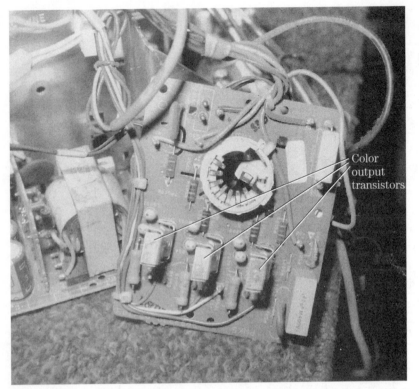

3-8 The three color amps are mounted on the CRT PC board.

3-9 Here, one large IC contains most of the circuits.

3-10 The present-day TV chassis looks rather small in a TV console cabinet.

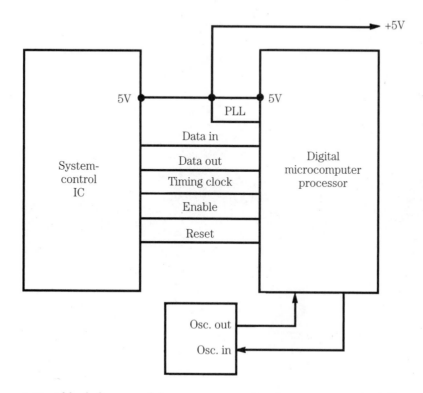

3-11 A block diagram of the system control in the more expensive TVs.

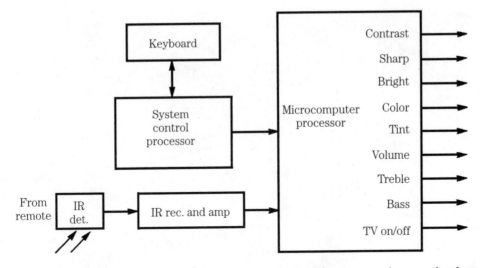

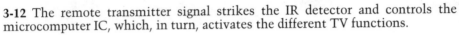

3-12 The remote transmitter signal strikes the IR detector and controls the microcomputer IC, which, in turn, activates the different TV functions.

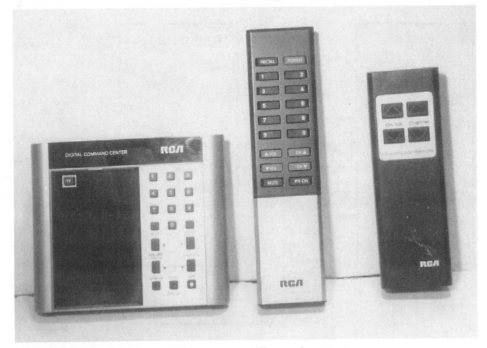

3-13 Today's remote controls have many different functions.

Comb filter

The comb filter is an electronic device to separate luminance (brightness) and chroma (color) video signals, which eliminates the cross-color or rainbow effect in the picture. The signal from the video stages applies at the input of comb filter network and feeds to the luminance (brightness) amp and restoration circuits. A time clock signal from the luminance/chroma processor is applied to the comb filter charge-coupled device (CCD). The CCD is actually a time-delay line. The comb filter is only used in a few color TVs (Fig. 3-14).

Color picture tube

To better understand the construction and operation of a color picture tube, quickly review the features of a monochrome picture tube.

A monochrome TV picture tube consists of three main sections: the gun, the deflection system, and the screen. The gun produces a stream of electrons, which, with the aid of properly applied voltages, is focused to a sharp point on the face of the tube (the screen). The deflection system, again by virtue of correctly applied voltages of proper characteristics (horizontal and vertical sweep voltages), moves the beam to the desired points, in proper sequence, on the screen.

Now, if instead of a black-and-white phosphor (coating on the inside surface of the TV screen), we substituted a red phosphor (or a blue, or a green), the same exact picture tube just described would produce the same TV picture, but in red (or blue, or green), with all the various graduations in brightness, as in the black-and-white picture. It is a bit of oversimplification, but nonetheless true, that the color TV picture tube is a combination of three tubes: a red tube, a blue tube, and a green tube in one glass envelope. However, although this picture tube has three distinct guns, each emitting an electron beam for each color, it has but one screen, for all

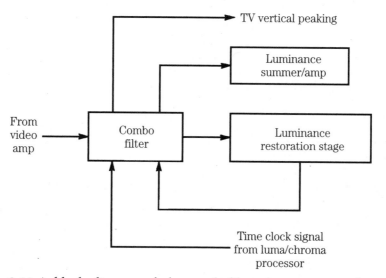

3-14 A block diagram of the comb filter that separates the brightness from the chroma signal.

three colors. This accounts for the great complexity of color TV tubes. To understand some vital adjustments and corrections that are sometimes necessary on a color TV, it is first necessary to understand the structure and functioning of a color picture tube.

In contrast with the black-and-white picture tube, which has a continuous coating of phosphor and where the electron beam can be placed on any point on the screen, the color tube screen does not have a continuous coating, but consists of a large number (hundreds of thousands) of color dots, and the electron beam must hit the dots only. Specifically, the screen is composed of three-dot groups; each group has a red, a blue, and a green dot closely spaced, but not quite touching. A recent improvement in color picture tubes has been realized by putting a black border (a nonluminous coating) around each color dot to reduce unintentional blending of the colors. The dot arrangement is alternated in each group so that no two adjacent dots in any direction are of the same color. For better visualization, consider the dot groups as subminiature billiard balls, arranged in triangles (three in a group, almost touching) with no two balls in the triangle being the same. Figure 3-15 shows the dot layout on a portion of a color screen.

Immediately behind the tricolor picture tube faceplate is a perforated metallic plate very precisely arranged and positioned so that each hole in the perforated plate lies in an exact position behind a group of color dots, a color triad, on the faceplate. When the plate is correctly positioned, the red beam goes through the hole and strikes the red dot only. Similarly, the blue and green beams passing through the same hole will strike the blue and green dots, respectively. This is a precision structural alignment, and it is independent of the characteristics of behavior of the TV set.

For the sake of clarity, the three segments of the electron gun in the tricolor picture tube are identical. They are called red, blue, and green only because of their physical positioning, with regard to the aperture mask (the plate behind the screen) and faceplate, not because of any color difference between them. All three produce identical electron beams with no inherent color characteristics.

It should be apparent from this description that for the proper excitation of the three colors the three electron beams originating at different positions in the neck of the tube (see Fig. 3-16) must be made to properly bend or converge so as to go

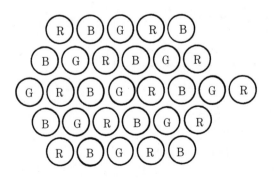

3-15 The dot pattern on the front of a color TV screen.

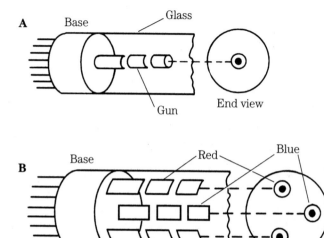

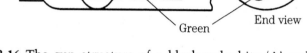

3-16 The gun structure of a black-and-white (A) and color picture tube (B).

through the holes in the aperture mask, instead of striking the space between the holes. Although this is approximately provided by virtue of tube design, it is not sufficiently accurate for a satisfactory color picture reproduction. To ensure this action, special convergence devices are used. There are two such sets of convergence adjustments. One consists of a set of magnets placed around the neck of the picture tube and adjustable for physical position. This set is called *static convergence controls* and is best left alone. The second set of convergence adjustments is of the rear-of-the-set types. Three groups of this type of adjustment controls exist, one for each of the three primary colors.

Although focusing is a function common to all TV picture tubes, it is much more important in a color tube because of the strict compatibility requirements between the three electron beams, the hole in the aperture mask and the color triad just in front of each hole.

Block S of Fig. 3-4 indicates the convergence function just described. Although the convergence adjustment shafts are accessible at the rear of the receiver, these adjustments require extreme care, as well as complex equipment to set them properly. Such is not the case with the focusing adjustments. With proper care, this adjustment can be satisfactorily performed by a beginner.

Incidentally, at least one color TV manufacturer has built in a convergence tests accessory, making it relatively simple to carry out these adjustments when required. The trend in the industry seems to be toward the inclusion of the simpler self-test or self-correct capabilities in the receiver. A common example is the demagnetization or degaussing coil, which prevents and corrects color deterioration because of stray magnetic influences in the immediate vicinity of the color set. Prior to this, the performance of a color TV set—even though it was in normal operating condition—was affected (sometimes seriously degraded) by the appliances or electric wiring in the immediate vicinity.

A third strictly color function mentioned earlier is the color killer, and it is described fully later. At this point, it is sufficient to say that its purpose is to ensure that no color appears on the screen during a black-and-white picture.

Picture tube elements

The internal elements of a picture tube includes a phosphor coating, accelerator grids G2, G3, and G5. G1 is the grid element with G4 as the focus grid. Three different color cathodes (red, green, and blue elements) are next to each heater element. The three different color guns have a separate heater or filament element wired in series with the other heaters. The magnetic deflection CRT requires a yoke mounted on the neck of the tube to deflect the electron beam (Fig. 3-17).

Magnetic deflection tubes might have screens smaller than one inch (camcorder) or larger than 35 inches diagonally. These tubes require a magnetic deflection yoke on the outside of the deflection area to move the beam up and down, and from left to right. Electrostatic deflection has deflection plates like that used in the oscilloscope CRT.

The filament or heater element heats up the cathode, which emits electrons. The incoming color signal is applied to each cathode element. The brightness of the beam is controlled by a negative voltage applied to the grid element (G1), compared to the cathode elements. If the voltage on the grid element is highly negative, no electrons will pass and no raster appears on the screen; the screen is black. Thus, the brightness control controls the voltage on the grid (G1), which, in turn, controls the brightness of the raster. The cutoff voltage is known at the exact point where electron flow stops.

G1 and G4 are usually tied together with the focus grid (G3) in between. G2 and G4 are the accelerator grids and G3 is the focus anode that pinpoints the electron beam.

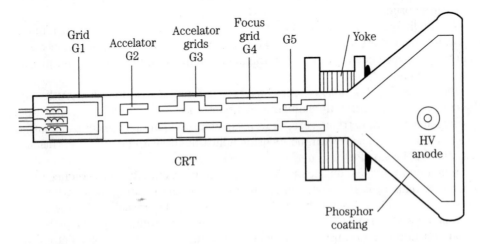

3-17 The internal construction of a color picture with the tube three-color gun assemblies.

The front part of the screen has a high potential voltage applied to pull the electrons from the cathode elements. The anode voltage of the very large picture tubes can be dangerous if more than 32 kV. The two vertical and horizontal coils deflect the beam. The polarity of voltage applied to these coils will deflect the beam up and down (vertically), and side to side (horizontally).

Be careful when working around picture tubes. Heavy tools or a loose TV cover could accidentally fall on the neck of the picture tube and break it off. Avoid the high-voltage anode connection. When measuring high voltage at the CRT anode button, be sure that the HV probe ground wire is clipped to the chassis ground or outside CRT ground wire. If not, you can receive a damaging shock when taking high-voltage measurements. Do not take a high-voltage measurement with the small DMM or VOM. Take high-voltage measurements with a high-voltage probe or a special meter.

CRT socket components

Many different components are connected or mounted on the CRT socket. Besides the focus cable, the color output transistors are mounted on a separate PC board attached to the picture tube socket (Fig. 3-18). Some color receivers have the different color bias transistors, which are directly coupled to the three color amp transistors. The red, green, and blue color output transistors are mounted upon the PC board (Fig. 3-19).

Usually, the voltage supplied to the color amp transistors is from the horizontal output transformer circuits. The color output signal is tied directly to the cathode terminals of the picture tube. The same voltage at the cathode terminal is about the same on the collector terminals of the color amp transistors (+200 volts). In this RCA portable, the focus and screen controls plug in after the CRT socket is in position (Fig. 3-20). If one color or all of one color on the picture tube is missing, suspect a defective color amp circuit.

Recent color tube improvements

Figure 3-15 shows the basic color tube using the three component colors in a triad arrangement, relying on the properly perforated plate or shadow mask behind the screen to converge the three component beams as required to produce the desired color. Even with a so-called border guard—a black, opaque area separating the color dots—and with improvements in convergence techniques, this construction left much to be desired from the viewpoint of color purity and picture fidelity. Subsequent to this, two significant improvements were made in color tubes so that now many TV sets have one or the other of these improved tubes.

The first of these was known commercially as the *Trinitron* produced by Sony Corporation. This uses a three-beam gun, with a color faceplate consisting not of triple dots, but of vertical strips. These strips are also separated by opaque strips to prevent unintended color overrun or blending. The strips are continuous from top to bottom. Figure 3-21 shows in sketch form, the gun, focusing, and faceplate structure of this tube. The Trinitron equivalent of the shadow mask, called the *aperture grille*, which has vertical slits instead of triad holes, is claimed to be more accurately producible and to provide a greater beam converage of the color screen. In a

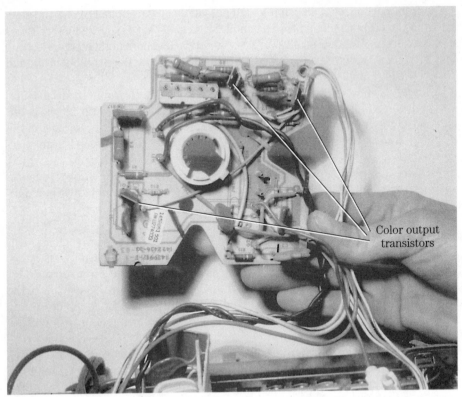

3-18 Locate the three color output transistors on the picture tube board.

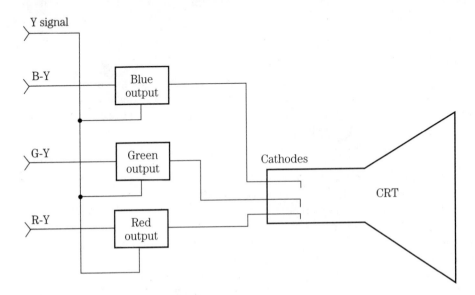

3-19 A block diagram of the color output transistors connected directly to the cathode elements of the CRT.

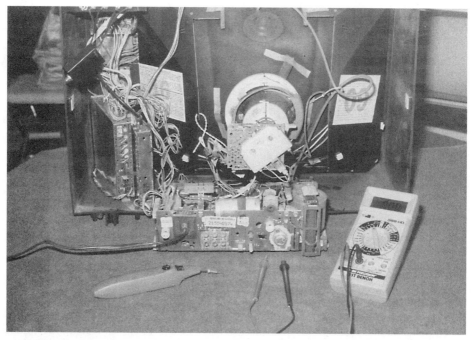

3-20 The focus and screen grid assemblies plug directly into the CRT board of the RCA CTC109 chassis.

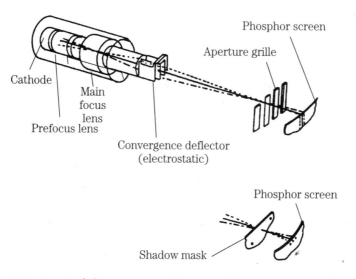

Phosphor screen

Aperture grille

Cathode

Main
focus
lens

Prefocus lens

Convergence deflector
(electrostatic)

Phosphor screen

Shadow mask

3-21 A typical construction of the Sony Trinitron picture tube.

properly adjusted set of this type, that is, when the brightness, contrast, and color intensity are normal, it is possible to actually see each of the vertical color lines by getting close to the screen. The appearance resembles that of a fine cloth weave. In addition to the superior picture quality claimed for this type of tube, it also has simplified convergence adjustments.

The second type of improved color tube, which is being increasingly used by a number of manufacturers, also has a vertical slot structure, but, unlike the Trinitron's continuous stripe arrangement, it consists of slots separated by opaque areas. In addition, these types use a three-gun in-line arrangement, in contrast to the triad gun arrangement and the single-gun three-beam type. It is claimed that this type of shadow mask also provides for better color distribution and depth. Both of these types, the Trinitron and the slotted-aperture type, are superior to the old, three-dot type of aperture structure. In addition, and perhaps only incidentally, these tubes are also shorter because of the wider angle encompassed by the screen. ALthough this might not be of particular interest to the TV owner, other than perhaps the decreased depth of the cabinet, there are some definite advantages from a technical viewpoint. Although there still are a few sets with the three-dot matrix type of picture tube, they are said to have been improved with regard to focusing and color depth.

There is no simple way to identify the tube type from casual viewing by the observer. But all the manufacturers describe their products so as to leave no doubt about these features. Thus, Trinitron is a sure identification of the continuous-strip mask type, *in-line* identifies the slotted-mask tube (other terms are used in advertising), and the improved three-dot types are identified by their manufacturers as the *mask focus*, suggesting that each mask hole acts as a focusing lens for sharper, brighter images.

One other feature of some of these newer picture tubes is worth mentioning, although it is technically of minor value. This is a mild light filter (picture tube seems dark when the set is off), of a smoky hue, intended as an antireflection device, so that the face of the tube will not act as a mirror and reflect objects in the room (Fig. 3-22).

The color tube has been significantly improvemed for the customer and service technician. The picture tube screens are much larger, have square corners, and do not need any type of convergence adjustments. RCA has developed a new color TV picture tube featuring a larger viewing area, flatter faceplate, and squarer corners than the current picture tubes.

Large picture tubes

Picture tubes are getting bigger every day. A few years ago, the 24-inch was the largest size with a 25-inch design around the corner. Since then, 26-, 27-, 30-, 31-, and 35-inch screens have been produced (Fig. 3-23). The 27-inch quite popular in portables and consoles.

The 27-, 30-, 31-, 32-, and 35-inch screens have a flat front surface with almost square corners. The picture tube is measured diagonally from corner to corner. You can see very little picture distortion in these larger glass picture tubes because of the pin-cushion circuits. Of course, the larger-sized tubes are a little more difficult

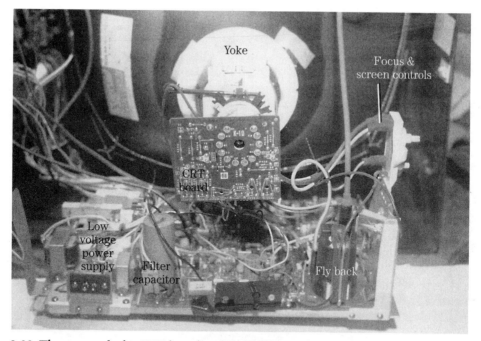

3-22 The rear end of a CRT board and the different parts on the TV chassis.

to replace. Two people are required to lift, remove, and replace the larger picture tubes. In fact, the picture tube replacement for the 35-inch CRT might cost more to replace than the whole TV receiver.

The defective CRT

How do you know the picture tube is defective or going bad? The defective picture tube will not light, appears whitish around figures, or has blotchy pictures and distorted colors. The picture tube heaters might open and not light at the end of the gun assembly (Fig. 3-24). When the picture appears dim with very little brightness, suspect a weak picture tube stream of electrons. The nickel-plated cathode around the heater or filament is worn off or bunches of electrons have built up on the emitting element.

The gassy picture tube might show blotchy areas with a close-up view of a person's face. Usually, the defects appear on the black-and-white picture. Extremely distorted areas of color when the chassis turned on indicates a defective CRT. Also, if the picture appears dim at first, then improves as the set becomes warm, indicates that the picture tube is defective. Some of the these symptoms can also be caused by a defective component in the picture tube circuits. Most picture tube testers will indicate if the CRT is defective.

HV CRT voltage measurement

Improper high voltage at the anode terminal button of the picture tube can have a low brightness, insufficient width, and distorted picture. Measure the high voltage

3-23 The 27-inch picture tube has square corners and a flat-front screen.

3-24 The defective CRT might have open filaments or heaters, a dim or weak picture with a weak gun assembly, and glassy-looking areas with signs of a gassy picture tube.

at the CRT to determine if the high-voltage circuits are normal with a black screen. Simply place the HV probe tip (preferably one of 42 kV) under the rubber anode cover and touch the anode button (Fig. 3-25).

Be sure that the ground test lead is clipped to the chassis or picture tube ground. If not, you can become shocked and probably drop the test probe. Be careful not to pry off the anode HV lead in the process. Excessive high voltage will began to arc over whatever object is close by. Read the measured high voltage directly from the meter probe.

On-screen display

One of the recent displays added to the many TV circuits is the on-screen display (OSD). The display of numbers, letters, and dots or dashes of the various operation controls are placed on the screen. On some TV screens, the contrast, color, tint, brightness, volume, treble, and bass bars or dashes, with correct function is shown on the face of the picture tube. You can adjust these controls manually or with the remote control. The operation is called *menus* or *characters* by some manufacturers.

By placing the function and graduation of bar dashes on the screen, you can see where you like the best-level color picture or sound. The dashes can go across the screen or when you select the best viewing picture for yourself, the color and brightness might be one-third the way up. This on-screen information is applied over the color telecast picture (Fig. 3-26).

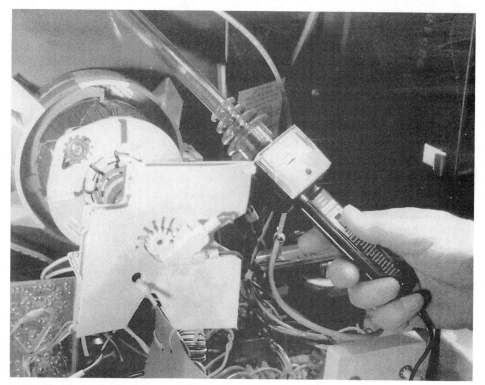

3-25 Check the high voltage at the CRT with a high-voltage probe.

3-26 The on-screen display can be controlled by the remote or manual operation.

The system control and microcomputer processor select and control the on-screen display process. The on-screen display is fed from the microcomputer to a processing IC. The OSD signal is demodulated and connected into the R-Y, G-Y, and B-Y CRT bias drive circuits (Fig. 3-27). A defective on-screen display might result in no or scrambled on-screen display; no green characters, a solid green screen; no luminance or brightness, but the color is normal; or no black edge. Troubleshooting the OSC circuits takes some know-how and an oscilloscope.

Picture in a picture

You might be able to see more than one picture on the picture tube at the same time. This process allows you to watch the large picture with one or more small pictures over the larger picture or vice versa. You can zoom the small picture in size and freeze it at that point, in some TV chassis (Fig. 3-28).

When the picture-in-a-picture function is selected, the small picture should appear in the lower right-hand corner of the screen. The on-screen display displays the big picture at the left and small pictures at the right. The small picture can be moved to any position on the screen by using the four direction arrow buttons. The picture will move until the arrow key or button is released. The sound is transmitted with the large or original picture on the screen.

There are many different functions found with the picture-in-a-picture process. These functions are controlled by a picture-in-a-picture module. Because many different signals and critical test equipment are required, the picture-in-a-picture module can be replaced or serviced properly by the electronic technician.

Big screen projection

Several large, curved-screen TV receivers on the market with front and rear projection. Three separate small-projection tubes with lenses are converged to the large projection screen. The control console with the projection tubes are mounted in front of the large, curved screen. With rear projection, all projection equipment is inside the TV cabinet. Most TV sets with rear projection have large, flat screens.A disadvantage of large-screen projection TV sets viewed in a normal living room setting was inadequate brightness and contrast. Today, projection TVs have improved brightness because of liquid cooling assemblies, which are a part of each projection tube: red, green, and blue. A cavity holding a cooling solution permits each projection tube to operate at higher voltage potentials, resulting in brighter, sharper pictures (Fig. 3-29). Each tube has its own cooling assembly, lenses, and control circuits.

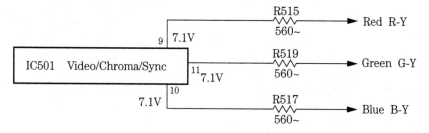
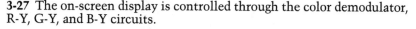

3-27 The on-screen display is controlled through the color demodulator, R-Y, G-Y, and B-Y circuits.

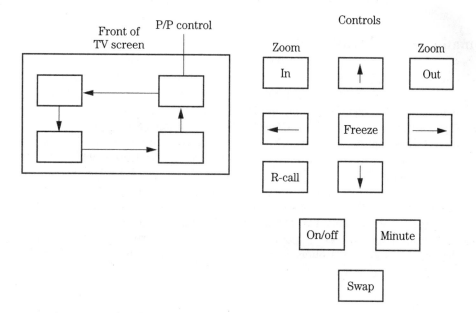

3-28 The "picture-in-a-picture" capability is used in the most expensive TV sets.

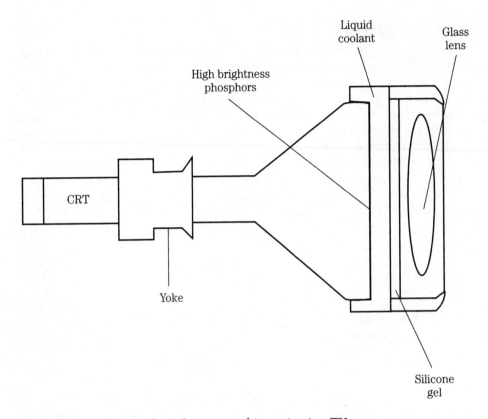

3-29 Three separate color tubes are used in projection TVs.

Small black-and-white TV screens

Several years ago, the smallest TV screen was 5 inches. The small screen size has now developed into the mini television screens from a 3-inch to a 2.7-inch and down to a 2-inch portable (Fig. 3-30). The pocket-type portable operates from flashlight cells, and the large-screen mini TV might be either ac or dc powered. You can find the small screen in a TV/AM/FM/cassette portable (Fig. 3-31).

Often, with small-size screens, the high voltage is around 1 kV or 1.5 kV at the anode connection. All circuits are solid-state with many ICs. Here is where the TV schematic is handy. In the early three-inch portable TV, the solid-state chassis was powered with a battery pack, which included several batteries or ac external power jack (Fig. 3-32). The small ac/dc portable TV can fit on a desk in the office or study. Here, the portable TV antenna dipole will pickup local TV stations (Fig. 3-33). The mini-screen TV components are packed and pressed close together, which demands a lot of patience, small, thin fingers, and a magnifying lamp to locate the tiny defective parts.

3-30 A pocket LCD screen TV for convenient viewing.

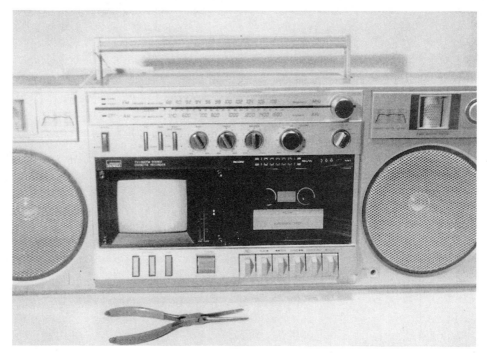

3-31 A small black-and-white TV in a boom-box cassette player.

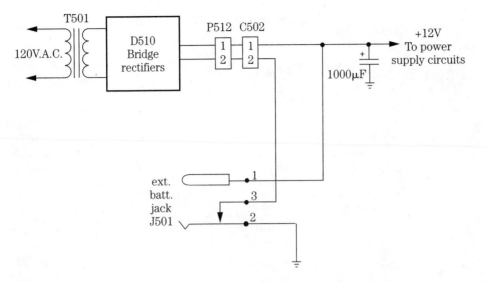

3-32 Battery and power-pack operation in a small-screen portable color TV.

3-33 The flat black-and-white or portable TV can operate from a single dipole antenna.

4
CHAPTER

Color-only sections

The video amplifier is the best place to begin examining the strictly color portions of a color receiver. That's where the differences begin. Figure 4-1 is a detailed block diagram of the color-only sections of a generalized TV set. It illustrates the functions, adjustments, and control locations of potential malfunctions that could be found in an actual TV set.

Beginning with block A, the video amplifier of the black-and-white portion of the set, the color signal together with its timing (burst or sync) signal go to block E on the color side of the dashed dividing line. The dashed line from block A to block F is an alternate way for the color timing signal to be taken out of the video amplifier; thus, in some sets, two separate signals might pass from the video amplifier to

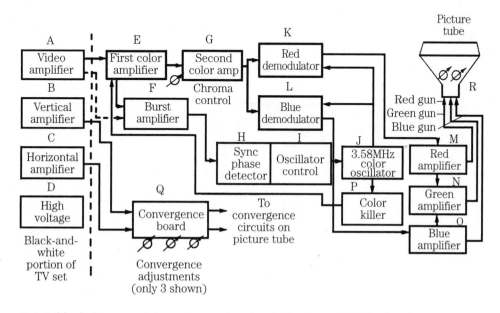

4-1 A block diagram of the various color circuits in the early TV color chassis.

the color portion of the set (the picture information to block E and the burst information to block F). In more recently manufactured TV sets, the combined signal is amplified in the first color amplifier, then the burst separates and goes to the burst amplifier (block F) while the color (chroma) signal goes to the second color amplifier, block G. Here again, some TV sets have only one color amplifier, so block G would be missing. Whether a particular TV set has one or two chroma amplifiers, the front-of-the-set control marked color is located here and it adjusts the color content of the picture from all black-and-white (no color) to full color (and beyond, if it is turned to its extreme position).

New color-processing circuits

In this particular color circuit, the video signal is applied to a 3.58-MHz bandpass filter (BPF). The video is separated from the color signal and only the color information is applied to pin 49 (Fig. 4-2). Test point TP201 indicates if the color waveform is present this far in the circuit. The color or chroma signal is located at the first and second amplifiers inside of IC1001. The gain of the first amp is controlled by the automatic color control (ACC).

TP205 is a test point to check the tint control voltage at pin 2. The horizontal blanking signal is at pin 13. Pin 16 is the supply voltage pin and it can be checked at test point TP204.

The gain at the second color amplifier is determined by the color contrast voltage on pin 3. The second color amp has a color-killer detector that kills the color when a black-and-white signal is received. The output of the second amp is connected internally to the chroma demodular circuits.

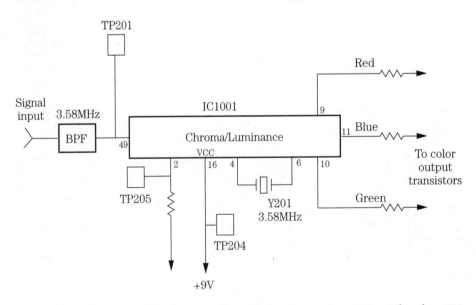

4-2 A block diagram of the latest color circuits in one large IC, with other TV circuits.

To detect the color signal, a 3.58-MHz (Y201) crystal is used. Y201 controls the oscillator frequency (VCO). The automatic phase control (APC) compares the phase of the oscillator to the burst signal and results in a phase error signal. This phase error signal keeps the oscillator locked to the burst signal.

The demodular signal is developed in the chroma demodulator circuits, R-Y, G-Y, and B-Y signals. These -Y signals are applied to each color output transistor. Check the waveforms at pins 9, 10, and 11 to determine if the color circuits are performing. Besides these waveform tests, check the 3.58-MHz oscillator signal at pin 6, color input at pin 49, and luminance waveform at output pin 13. All waveforms are taken with a scope while the supply voltage (V_{cc}) is tested with the DMM.

Burst amplifier

The simplest way to describe the function of a burst amplifier (block F in Fig. 4-3) is to say it is exactly the same as the function of the sync amplifier (block H, Fig. 3-4). The input to the burst amplifiers is derived either directly from the video amplifier (block A, dashed arrow) or via additional amplification in the first chroma amplifier (block E, solid arrow). The output of the burst amplifier is ultimately applied to the 3.58-MHz oscillator (block J) for precise control (timing) of the oscillator (Fig. 4-4).

Color sync and oscillator control

The color sync phase detector (block H) and oscillator control (block I) are grouped together for two reasons: first, both perform functions that are important to controlling the accuracy of the color oscillator, and second, they can be considered, in a sense, optional. Although some manufacturers use this function for oscillator control, others achieve the same result in a different way and do not incorporate these two in the TV receiver design. What is important to the TV set owner is that when these functions are in a set, they are links in a chain. If one or both should fail, the 3.58-MHz oscillator might wander sufficiently to cause some serious deficiencies in the color rendition and balances (by cutting off some of the color frequency components in the output signals).

The 3.58-MHz oscillator is one of the ingredients in the conversion of the chroma signal to the red, blue, and green portions of the final color picture. The balance between these three primary colors is, of course, essential for a normal color picture and can be achieved only when the 3.58-MHz oscillator signal is fed to the demodulators exactly in time with the corresponding signal in the transmitter. Control circuits and tubes in blocks H and I establish and maintain synchronization between the transmitter and the receiver. A failure here invariably degrades, if not altogether destroys, the normal color picture.

Color oscillator

The 3.58-MHz precision oscillator (block J, Fig. 4-1) is a crystal-controlled oscillator, which is basically very accurate and stable—even before being further controlled by the burst signal from the transmitter. An examination of the immediate vicinity will locate the crystal. It is a fragile, tissue-thin wafer of a synthetic, quartz-like material (the natural mineral quartz is seldom used nowadays) encased in a hermetically sealed plug-in unit, as illustrated in Fig. 4-4. Although the little box

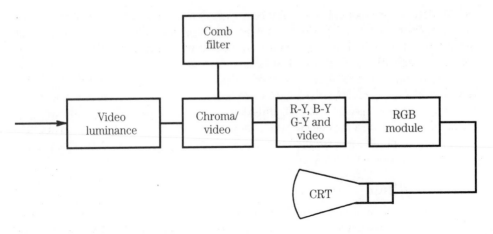

4-3 A block diagram of the chroma circuits.

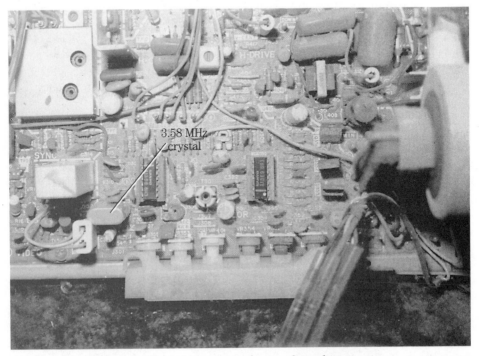

4-4 The 3.58-MHz crystal can help you to locate the color circuits.

requires careful handling, it is relatively durable; however, occasionally it might become intermittent (stop functioning, then start again after the TV set is switched off and on) or it can fail completely. In either case, if the crystal malfunctions, the color, but not the picture, disappears.

Solid-state 3.58-MHz color oscillator

Within the solid-state circuits, the 3.58-MHz color oscillator can be located in a transistor or an IC. When the IC chip was introduced, the color oscillator was included into the video-luminance-chroma IC (Fig. 4-5). Today, the chroma processing circuits are included with the luminance and chroma processor. You might find the chroma sections in one large module that is bolted or snapped into the chassis (Fig. 4-6).

The chroma signal is demodulated with a 3.58-MHz voltage-controlled oscillator (VCO). The frequency of the oscillator is controlled or stabilized by a fixed crystal (Fig. 4-7). The automatic phase control compares the phase of 3.58-MHz oscillator signal to the burst signal and provides an error signal. This error signal locks the phase of 3.58-MHz oscillator to the burst signal.

Color demodulators

Blocks K and L are the color demodulators. As shown in Fig. 4-1, the combined color signal comes from the (second) chroma (color) amplifier, and the second input is from the precision-controlled 3.58-MHz oscillator (or reference signal). Notice that there are only two demodulators, one for the blue signal, the other for the red. The green demodulator, as such, does not exist, and it is unnecessary. Because a combination of the primary colors, blue and red, will produce green, no separate demodulator is required. The green signal is derived by mixing a proportional amount of red and blue.

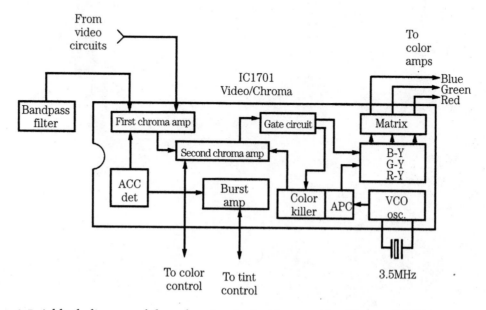

4-5 A block diagram of the color circuits inside one video/chroma IC1701.

4-6 One large IC with the 3.58-kHz crystal nearby helps you to locate the color circuits.

4-7 The location of the 3.58-MHz crystal in a recent TV.

With today's chroma demodulator circuits, all three colors are detected in one circuit inside a large IC. The function of the modular stage is to detect the color information from the combined color and carrier signal. The output of the color demodulator stage feeds color signal to each individual color amplifier.

The color waveform on each output terminal is quite similar on the oscilloscope. Most color circuits are signal traced and serviced with a color dot-bar generator connected to the TV antenna terminals. The color signal is sent to each base terminal of the color output transistor, and the luminance or video signal is applied to the emitter terminals (Fig. 4-8).

Solid-state color demodulators

The outputs of the chroma limiter and phase detector are applied to the color modulator circuits. Sometimes the phase detector is called a *flesh* or *hue detector*. The output of the modulator is applied to a buffer stage with the generator 3.58-MHz voltage-controlled oscillator. The output of the buffer stage, which is a tint-controlled signal, is applied to the phase-shifting components.

The output signals from the phase-shifting circuits are applied to the chroma matrix amp. The chroma matrix amp produces R-Y, A-Y, and B-Y signals combined with the luminance signal. R, G, and B output signals are fed from the chroma processor circuits to the respective bias and color amplifiers (Fig. 4-9).

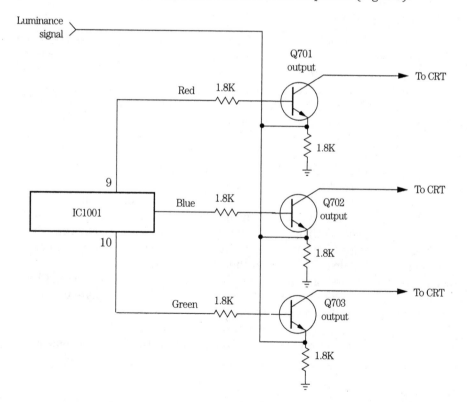

4-8 The color output transistors are located between the color output demodulator and the picture tube.

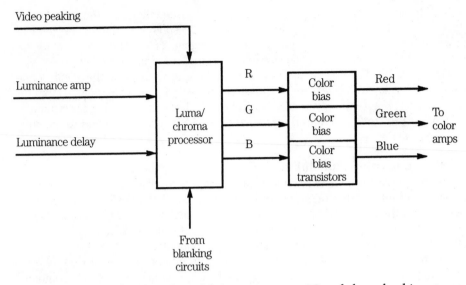

Video peaking

Luminance amp

Luminance delay

Luma/chroma processor

From blanking circuits

R   G   B

Color bias

Color bias

Color bias transistors

Red

Green

Blue

To color amps

4-9 A block diagram of the luma/chroma processor IC and the color-bias stages.

Color amplifiers

In some color TV circuits, red, green, and blue bias transistors are used between the color demodulator terminals and color output transistors. Each transistor has a bias drive resistor in the emitter or luminance input circuits (Fig.4-10). Here, separate color amplifiers are connected through an isolation resistor to the emitter circuits. The color output amplifier signal is fed from each collector to the cathode terminals of the picture tube. The color signal is applied to the base terminal of each bias transistor, and the luminance or video signal is connected to the emitter terminals of the bias transistors.

Today's color amplifiers are fed directly from the demodulator pin terminals, through the isolation resistor, and to the base terminal of the color output transistors. The chroma signal is applied to the base terminals of the output transistors with the luminance signal to the emitter terminals (Fig. 4-11). The color output amps have a very high voltage on the collector terminals and might normally run a little warm. These color amps often have a waveform of 2 volts (at the base terminals), which is then amplified up to 100 volts at the collector terminals.

When one color is missing from the raster, suspect a color amp transistor. Check the picture tube with a color tube tester to be sure that the color gun assembly is not defective. Next, measure the collector voltage at each color output amp. Low voltage at the color amp can indicate a leaky color output transistor, if the plate load resistor is quite hot. Usually, when a problem occurs, the color amp transistors become leaky or appear to be open.

Color killer

Block P represents the color killer, a violent title that means just what it says. The function of this section is to prevent any color from showing on the screen when

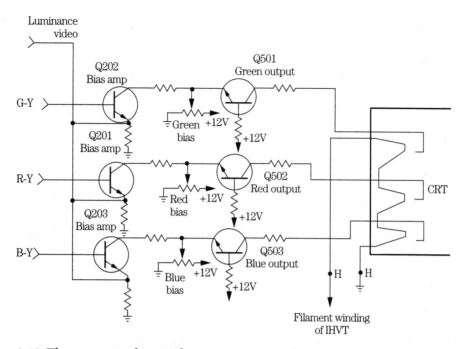

4-10 The separate bias color transistors and color output transistors connected to the cathodes of picture tube.

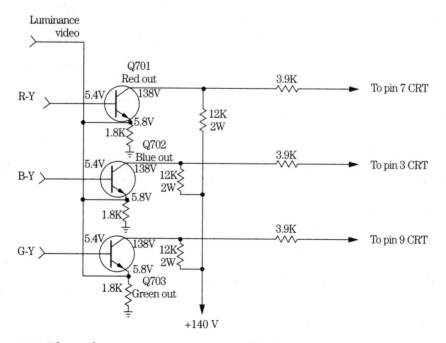

4-11 The color output transistors with corresponding emitter and collector load resistors.

a black-and-white picture is being received. To begin with, remember that a monochrome picture in a color TV set is produced from the correct combination of all three primary colors. In other words, a black-and-white picture on the screen comes not from no color, but from all colors in the exact proportion, just as white is not the absence of color, but the presence of all colors.

Second, and this follows, to transmit a black-and-white picture requires proper balance of all three primary color components. Although balance would ensure no coloration of a black-and-white picture, such balance cannot be taken for granted. Consequently, color fringes might appear on the color TV set when the picture is transmitted in monochrome.

Remember that in all cases, whether monochrome or color pictures are transmitted, the picture tube receives a brightness signal, which is actually the picture minus the color. Of course, when no color is transmitted, the chroma amplifiers have nothing to contribute. Similarly, when a color picture is received on a black-and-white set (in which there obviously are no color amplifiers), this brightness signal is all there is and it produces a black-and-white picture without any special adjustments or color killers.

Finally, the color amplifiers (as well as the red, blue, and green guns of the picture tube) operate on current (electron) beams, not on beams of colored light; thus, any current in the path of a chroma amplifier (which automatically feeds one of the three color guns) is capable of exciting the color dots on the picture tube screen, regardless of whether or not color is being transmitted. To prevent this unwanted color, the color-killer function is required.

In operation, the color-killer circuit keeps the chroma amplifier in a nonamplifying condition (or cutoff) until a color picture is received. At such time, a signal from the color sync (burst signal) disables the color killer, releasing the chroma amplifiers to perform their normal functions.

Improper adjustment of the color killer and fine tuning knob can result in loss of color or intermittent color in the picture. Sometimes the manual fine-tuning knob can be set clear to one side with the AFT button (auto fine tune) in operation. When the AFT button is switched off, the color might disappear. First, unlock the AFT button, then readjust the fine-tuning knob for normal picture and sound. If there is no color in the picture, turn the color control to its maximum position. Then locate the color-killer control (often located on the back chassis panel), turn it to the extreme clockwise position, then back counterclockwise. Watch for the color to return. Some color-killer controls give better color to the left or right side of rotation. Adjust both the color killer and color control (on front of TV) for the best color picture. Now, readjust the fine-tuning control and lock the color in with the AFT and color-monitor buttons.

Solid-state automatic chroma control (ACC)

The ACC circuits stabilize the chroma output signal against any fluctuation of the chroma burst signal. A voltage applied to pin 5 controls the gain of the first chroma amplifier (Fig. 4-12). When the burst signal at pin 1 and burst output signal is quite large, the voltage at pin 5 lowers, reducing the gain of the first chroma amplifier. R805 and C805 are the ACC filter network.

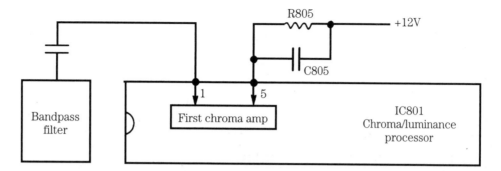

4-12 The block diagram of the solid-state automatic color control (ACC) circuits.

Solid-state color-killer circuits

The color-killer circuit is used to prevent colored noise, which appears as colored snow, from appearing on the picture tube when a black-and-white telecast is received. In the early tube and transistor chassis, a separate color-killer control was used (Fig. 4-13). Now the color-killer circuits operate automatically inside the chroma processing section. No color-killer controls are used in the latest TV chassis.

When no burst signals are applied from the color processor, the color-killer circuit output voltage lowers, making the gain of the second bandpass amplifier zero. When a burst signal is applied, the color-killer circuit is off, resulting in a normal color picture.

Within the present-day color-processing circuits, when a black-and-white signal with poor or weak color information is received, the color-killer circuit activates, turning the second chroma amplifier off. Then, when a good color signal is picked up, the color killer is turned off, and the output signal from the second chroma amplifier passes to the demodulator circuits (Fig. 4-14).

Solid-state automatic phase control (APC)

The APC circuits regulate the subcarrier oscillator circuit to maintain proper oscillator frequency and phase to ensure correct color synchronization. The automatic phase-control circuits detect the phase difference between burst signal from the burst amp and 3.58-MHz oscillator frequency.

The APC adjustment can be preset, so field adjustment is not required. The pot can be sealed, but the seal can be broken for APC adjustment, if required.

Within the latest chroma-processing circuits, the APC circuit provides automatic search function and is activated when the color killer is turned on (Fig. 4-15). The APC circuit is supplied to the variable-control oscillator (VCO) and tint-control circuits. The output of the tint control is applied to the chroma demodulator and R-Y, G-Y, and B-Y signals. These color circuits might be located within one large front end RF/IF luminance/chroma IC processor (Fig. 4-16).

IC chroma circuits

One large IC might include AGC, IF, AFT, IF sound, audio amp, sync separator, vertical and horizontal deflection, X-ray protector, and luma/chroma processing. The

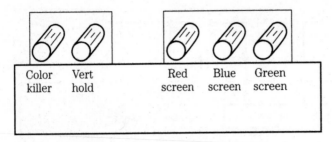
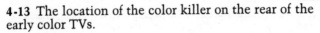

4-13 The location of the color killer on the rear of the early color TVs.

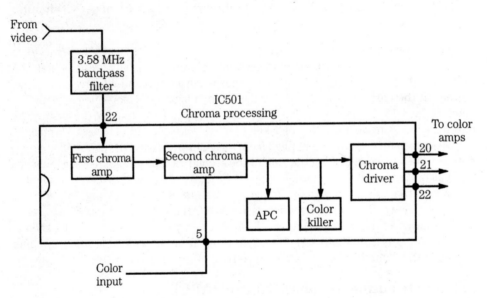

4-14 A block diagram of the automatic color-killer control circuit in a recent TV.

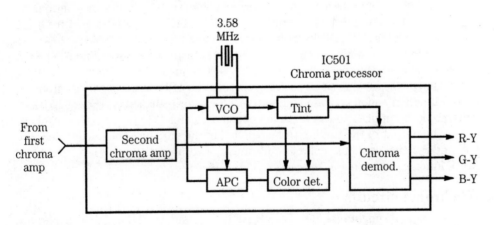

4-15 Today's APC circuits, which control the correct color synchronization of the color-processing IC501.

Large IC

4-16 One large IC might contain IF/SIF/video/AGC/sync/chroma/demodulator circuits.

IC color-processing circuits consist of first and second color amps, APC, VCO, color-killer detector, tint, and chroma demodulators (Fig. 4-17). The early solid-state chassis might contain these circuits on one large board (Fig. 4-18).

The video signal is applied to the 3.58-MHz bandpass filter. The bandpass filter passes only color information. This color signal is applied to the first and second color amps. The gain of the first color amp is controlled by the automatic color-control circuits (ACC). The control voltage is applied to the ACC detector to track the color and luma level.

The output of the first chroma amp is applied to the second color amp. The gain of the second color amp is controlled by the color-control voltage. The color-killer detector connects to the second color amp. When a black-and-white picture signal is received, the color-killer circuit turns off the second color amp. But, when a clean color is received, the color killer is turned off, letting the color signal pass to the color demodulator circuit.

The chroma signal is demodulated with the 3.58-MHz voltage-controlled oscillator (VCO). Then the automatic phase control (APC) compares the phase of the VCO and color burst signal, resulting in an error-phase signal. This error signal keeps the VCO locked to the burst signal.

The chroma demodulator signals are connected to the bias and color amp CRT drivers. The demodulated signals can be signal traced with a color-bar generator and scope. Many of the latest TV chassis have the color amps located on a separate CRT board (Fig. 4-19).

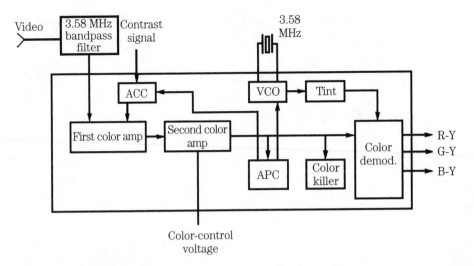

4-17 Here, the chroma-processing circuits are located with other IC circuits.

4-18 The location of the color circuits in a late-model color portable TV.

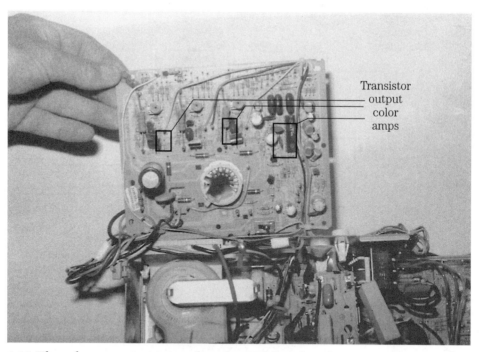

4-19 The color amps are mounted on the CRT board in the recent TV circuits.

Solid-state CRT bias and driver color circuits

The solid-state bias and driver amps consist of separate transistors that apply luminance and color to the picture tube. The G-Y, R-Y, and B-Y signals from the chroma processor are applied to each color-bias transistor. The luminance signal is applied to the emitter terminals of the green, red, and blue bias transistors (Fig. 4-20). This allows the luminance information to be applied to the demodulator chroma signals.

The color amp drivers are operated in a common-base circuit. The green, blue, and red bias voltages are applied to the emitter and the output color drive signal from the collector terminal applied directly to the respective color cathode in the color CRT gun assembly. Suspect the respective color driver amp transistor when one color is missing or if the CRT is defective. A leaky color amp might make the face of the picture tube have a dominant color.

Convergence

Convergence (block Q) is mostly a problem with three-gun picture tubes. In the old-fashioned black-and-white picture tube (and in the Trinitron color tube), the problem of convergence is greatly simplified. Convergence means aiming two or more beams so that they converge and meet at one and the same spot. As will be recalled from the description of the three-gun color tube, the aperture mask behind the color screen has one pinhole behind each group of three color dots (triads). Not only must the three electron beams go through the same pinhole, but they must be so positioned and directed that the red beam, after passing through the aperture,

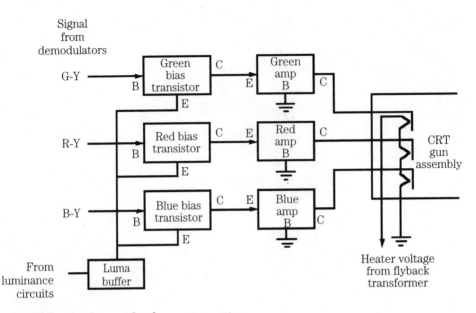

4-20 The color-bias and color-amp transistors.

strikes only the red color dot, and not one of the other two. Similarly, the blue or green gun must be so oriented that only the blue or the green dot is excited. The critical function of the convergence circuits and adjustments is to ensure this if true color separation is to be realized.

Notice that the basic compatibility between the three color guns, the single shadow mask aperture, and the three color dots just in front of this aperture is ensured by the original construction of color tubes. This, however, does not obviate "live" in-the-receiver adjustments and corrections.

There are two types of convergence adjustments in current three-gun color TV sets. One is called *static convergence*, the other dynamic convergence.

Static convergence

Static convergence of the three beams is a mechanical adjustment of the positions of small magnets on the neck of the picture tube. Its purpose is to have the three beams converge at the center of the screen only. Stated simply, the purpose of static convergence is to ensure that the three guns, starting (necessarily) from three different positions at the back end of the tube, produce one and only one spot when pointed to the center of the screen. This is the correct starting point for dynamic convergence.

Dynamic convergence

The need for dynamic convergence stems from the fact that the three beams travel different (longer) distances when going to the far corners (edges) of the tube than when going to the center of the screen. Figure 4-21 shows what a crosshatch pattern would look like as a result of the difference in the paths the beams travel.

Furthermore, because the length of the path from the gun to the screen is different for each spot on the screen (the shortest being to the center, the longest to any of the corners), it is necessary to continuously adjust the convergence forces as the beams sweep the screen. This is accomplished by dynamic convergence, continuous convergence correction during sweep.

In most color TV sets, all dynamic convergence circuits and adjustments are grouped together, usually on a discrete circuit board, often called the *convergence board*. Although the required continuous correction occurs electronically, it is first necessary to preset a number of controls (two or more for each color) so that the automatic electronic correction will be just right. In other words, dynamic convergence adjustments are intended to preset initial conditions from which the automatic circuitry takes over.

Because of the complexity of convergence adjustments, the procedure is not within the ability of the majority of beginners. This is so much caused by the technical difficulty as it is to the need for some very specialized test equipment. At least one TV manufacturer, however, incorporates such "specialized test equipment" into its TV sets and provides lucid instructions for convergence adjustments. But, even for the majority who will not be able to perform this function themselves, understanding the whys and hows will help them to understand what a professional TV servicer is doing.

To illustrate the phenomenon of nonconvergence, Fig. 4-21 shows a crosshatch pattern. In Fig. 4-21A, the lines are essentially straight to the very ends. In Fig. 4-21B, noticeable curvature appears and increases the farther the line moves from the center, in both the horizontal and vertical directions. In an actual color, picture improper convergence is accompanied by some color distortion and fuzzing near the ends of the lines. But for proper examination and correction, a test instrument capable of producing a crosshatch pattern on the screen, as in Fig. 4-21, is required. Such is the specialized test equipment referred to earlier.

The convergence procedure involves a sequence of adjustments of convergence board controls while observing a crosshatch pattern (from the test set) on the screen. Each color and each position (left, right, top, and bottom) are individually adjusted until the overall crosshatch is as linear as possible. Some sets have two sets of controls, one for blue, the other for red and green combined.

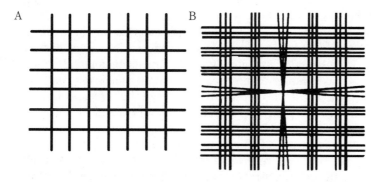

4-21 The crosshatch pattern (A) shows ideal convergence. Pattern (B) shows poor convergence.

Convergence problems are neither frequent nor chronic. They might occur when a picture tube is replaced, but seldom otherwise; however, should the need be visually apparent, you know what is needed. I mentioned earlier that one of the color TV sets on the market has such a color signal generator built in. The adjustment procedures for this set are quite specific, as they should be. They serve a particular set and can point to each control by symbol and location, and outline step-by-step procedures for the complete convergence adjustments for this particular set only.

Fixed static convergence

For some time now, improvements in picture-tube design, as well as in deflection techniques have made static convergence obsolete so to speak; thus, the starting point for convergence adjustments is the dynamic type—no static adjustments are required. In addition, dynamic convergence has been simplified. Because these changes are very much peculiar to each set, no general procedures can be given; the service information for each particular model should be followed for detailed procedures.

Color picture-tube circuits

Block R in this typical color chassis (Fig. 4-1) contains the color picture tube and its associated controls. There are a number of adjustments and controls, usually two for each color gun, although the red gun might sometimes have only one. Other adjustments are associated with the picture tube, but these are on the yoke and other assemblies on the tube neck, which is covered later.

At this point, be concerned with the controls relating to the color aspects of the TV set only. The two groups of adjustments are the screen and drive adjustments. In most receivers, there is a screen adjustment for each color—red screen, blue screen, and green screen; however, there might be only two drive adjustments, one for blue and one for green. The design of the more-recently manufactured color tubes and receivers often eliminates the need for the red drive adjustment.

Although these controls, whether six or five (or even fewer in some sets), are marked with particular color names, their adjustments are best made for a balanced black-and-white picture. These adjustments are often referred to as the *gray-scale adjustments*, meaning that they are set so that the complete range or scale of light is reproduced on a monochrome scene. Once this is done, the color picture should require little, if any, touching up for the best color balance.

The basic need for these adjustments stems from the fact that the three-gun picture tube contains three fairly independent tubes with some variations in characteristics between them. Adjustments are intended to compensate for and equalize these differences, as well as for some differences in the three signals reaching these three guns. Furthermore, because the correct proportions of the three colors are required to achieve a pure black-and-white picture, correct adjustment for such a picture, therefore, implies that the color balance is correct.

Screen and drive adjustments, although interdependent, can nevertheless be made one set at a time. First, the receiver brightness control is set for a fairly middle-to-dim level with the color control to minimum. Next, each screen control is advanced to the point where a trace of color appears, then is backed off to a point just beyond where this trace of color disappears. Next, the brightness control is advanced to what is considered normal brightness and the drive controls adjusted for a full gradation from dark to white on the screen, without any black or white patches showing. In other words, the complete range of light to dark should exist. A second touch up of the screen controls might be required, depending on the end results.

If this is carefully carried out, there should be good color balance when a color picture is received. A balanced color picture is what seems to be the most natural.

Perhaps it might help to indicate what is not a balanced color picture. If, for example, a color picture has a bluish hue (or greenish, or reddish), regardless of whether it is sky, or skin, or grass, obviously there is an excess of blue (or green, or red, as the case may be) in the picture. It is actually possible to create such an artificially tinted picture by advancing one of the screen controls.

Be cautious, however. If you want to make the test, you can best ensure restoration of the color balance to its original state by observing, and perhaps marking, the position of the control about to be changed. Then it is only necessary to reset the control to its original position to obtain the original color balance.

For best overall results, the proportion of the three colors is such that all white or all blue or all green are easiest to reproduce, but gradations and shadings are not so easy to reproduce accurately. The eye is very tolerant on color shades, except on such tints as flesh tones. Although it is possible to adjust the color balance of a set to favor flesh tones, such favoring is usually achieved at the expense of color balance; that is, the general background (or some not so intended objects) will take on the characteristic flesh tones. The clue to the problem lies in the favoring. If for technical (design transmission, etc.) reasons, flesh tones do not look natural on a properly adjusted color receiver, misadjusting the color balance circuits to favor the flesh tones will invariably distort the color balance on most other tints. It is like misadjusting the tone control on a radio receiver or phonograph to obtain better bass (low frequency) response by cutting off the high and medium frequencies, which are essential to balanced sound.

Another form of color imbalance is poor purity, which produces tinted raster of various colors in a black-and-white picture and incorrect hues in a color picture. Most cases of poor purity are caused by the TV's faceplate becoming magnetized from stray external magnetic fields. Stereo speakers adjacent to the TV cabinet, small radios or stereo units on top of the TV set, and turning off the vacuum cleaner next to the TV screen are some possible sources of external magnetic fields. In the majority of cases, poor purity can be corrected without removing the back cover of the TV, by degaussing (a procedure using a coil designed to demagnetize) the TV faceplate. Figure 4-22 shows a cluster of magnets around the neck of the picture tube; they are the purity magnetic assembly. These round magnets are locked together and preset; do not adjust or move them unless you have the knowledge and test equipment to do so.

Color dot-bar generator

The color dot-bar generator is ideal in setting up the picture-tube adjustments, leveling the deflection yoke, pin-cushion adjustments and servicing the chroma circuits. A low-priced dot-bar generator can be purchased for less than $150 (Fig. 4-23). Actually, the chroma circuits cannot be scoped successfully unless a color generator

4-22 Notice the various magnets and adjustments on the neck of CRT for convergence and purity.

4-23 The low-cost color dot-bar generator is required to service the color circuits.

is attached to the antenna terminals. It is most difficult to service the color circuits with the broadcast color signal.

The color generator can be used when adjusting the color dots on the TV screen and in convergence adjustments. Rotate the color generator to a cross-hatch pattern and make width, vertical linearity, and a level picture. If the picture is tilted to one side, loosen the yoke assembly and rotate the yoke assembly to level the picture (Fig. 4-24). Be careful not to throw out the convergence or purity adjustments. Touch up the purity and convergence, if needed.

4-24 Use the crosshatch pattern to level the picture and check vertical linearity on the TV screen.

5
CHAPTER

Introduction to troubleshooting

It is not my intention to tell you what to do without also explaining why something is to be done, or why a certain procedure should be followed. Unless you first understand what a tube, a transistor, or an assembly is supposed to do, basically, your chance of repair success is much smaller than it could be. That is also what I mean by potentially unsuccessful fixes.

Although it is possible for you to look on a chart of symptoms and find the one that seems most like that in your TV set, follow the suggested remedy, and actually get the desired result, the likelihood of such luck is rather small. However, your luck will improve in direct proportion to your familiarity with your set and your understanding of how it functions when operating normally. It is not impossible for you to look up a fault in the index, which will refer you to a certain page of this book, and, by following the instructions, end up with a fixed set. But your chances of continued success are better if you thoroughly acquaint yourself with the material in the first four chapters before tearing into your set.

Equipment

Although you should not buy a set of professional instruments and tools just to make some simple repairs, a few are necessary (in addition to the usual simple tools, such as screwdrivers, pliers, etc.).

Multimeters

For checking the continuity of antenna cables, speakers, tube filaments, and coils, select a low-priced digital multimeter (DMM) or volt-ohmmeter (VOM). You can find many uses for these small, low-priced test instruments around the house. Besides taking continuity tests, you can check resistance, voltage, and current measurements. With the latest DMMs, you can even check transistors, diodes, and capacitors (Fig. 5-1).

5-1 The analog VOM or digital multimeter (DMM) is handy when taking tests on the TV.

The typical DMM has ranges of 0 to 1500 Vdc, 0 to 100 Vac, ac or dc current from 200 μA to 2 A, 200 to 20 MΩ, and it also contains a diode tester. The diode test can be used to check a leaky or open diode, or make transistor-junction tests. The transistor-junction test indicates if the transistor is leaky, open, or has a high internal-resistance junction.

The typical low-priced VOM covers the following ranges: 0 to 10 kΩ, 0 to 1200 Vac and Vdc, and a dc current range of 60 μA to 300 mA. The VOM might load down the circuit being measured while the DMM does not. The DMM can measure a fraction of a volt, and the VOM is not that accurate. So, if you are going to purchase a new test instrument, make it a pocket DMM.

The DMM might also contain a frequency counter, capacitance tester, logic tester, transistor tester, diode tester, and an audible continuity alarm. The frequency-counter capability could extend up to 200 kHz with a resolution of up to 1 Hz. The capacitance range might extend up to 20 μF. Bipolar transistors can be tested for H_{FE} gain from 0 to 1000. The audible signal can be used with the diode and continuity tests.

Transistor/semiconductor tester

A good transistor, FET, SCR, and triac tester can save a lot of valuable time servicing a TV. You can check the semiconductors in and out of the circuit. The transistor tester can identify a npn or pnp transistor, an FET gate lead, and all transistor leads. Most transistor testers have a good and bad test.

You can purchase a transistor tester kit for less than \$35 (Fig. 5-2). The Elenco E6-DTK100 transistor tester kit is a nice one. The small diode-transistor kit will test all germanium, silicon, power LEDs, and zener diodes. The diode test will also indicate the cathode and anode leads of the diodes. Various germanium, silicon, power, RF, audio, switching, and FET transistors can be tested. The transistor tester can identify npn and pnp transistors. In-circuit transistor tests can be made with base and collector resistors low as 100 Ω.

Diode tests

Place the diode switch in the Diode position and connect the diode to the red and black test leads. Push in the red cotton test switch. One diode LED should blink no matter how the test leads are connected. If both diode LEDs blink, the diode is leaky or shorted. When neither LED lights, the diode is open.

Transistor tests

For out-of-circuit tests, place the transistor in the blue socket or attach leads to the C, B, and E leads. The green clip is the collector (C), the base (B) terminal is the yellow lead, and the emitter (E) terminal is the black lead. Push in the test button. Adjust the base control so that the OK red LED lights. This indicates that the transistor is normal. If the OK lamp does not light, turn the control up so that either the np or pn LED lights, indicating if it is a npn (np) or pnp (pn) transistor.

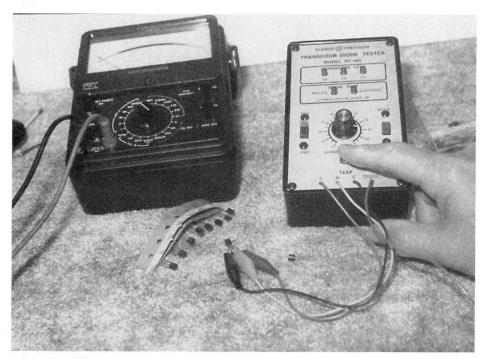

5-2 A low-priced transistor tester can help you to weed out those defective transistors, diodes, SCRs, and triacs.

If no LEDs light, the transistor might be open or the correct base terminal might not be connected. If the OK LED lights, the base indicates that it is a beta transistor. The lower the base control settings, the higher the beta.

In-circuit testing of a transistor can be performed in the same manner. Remember that a leakage test might be indicated if a resistor, coil, or components under 100 Ω are found in the base or collector terminals. Remove the base of the transistor from the circuit and take another test.

Hand tools

The first of these tools is a small soldering tool, about 100-watt capacity, preferably one of the instant-heating types (commonly called a *soldering gun*). This is not to suggest that you immediately attempt to unsolder and solder parts of your TV set. It is rather a form of insurance in case of accidents. For example, if, during a simple adjustment, you unintentionally break an interconnecting lead (a number of these seem to be floating around the back of every TV set behind the cover), you should be able to repair it without having to run for help. With a soldering gun, you need some solder (rosin core only, never acid core or plumber's solder), preferably of the "thin-wire" type. A half-pound spool of this stuff costs a couple of dollars or so from an electronics store and will last almost a lifetime (unless you are in the business full time).

The second item of great help to any do-it-yourselfer is a schematic diagram or set of diagrams and other service data for your set. This can sometimes be obtained by writing to the manufacturer, giving the model of the set and the chassis number. This information is always printed or stamped on the back cover of the set or the back apron of the chassis. Sometimes the chassis identification is also listed on the transistor layout chart, usually located inside the cabinet.

There is another and much more readily available source of service information. Instead of writing to the manufacturer, schematics, layouts, adjustment instructions and other helpful hints can be bought in almost any radio supply house. Although most users of this book might not be familiar with schematic diagrams, these nevertheless are very useful because they provide transistor identification numbers, their functions, as well as their relative physical locations in the chassis.

The last item of equipment is a substitute electric cord and plug, known as a *cheater cord*. The name is appropriate because it enables its user to cheat (defeat) the protective system designed into nearly all TV sets, namely prevention of the uninformed person from reaching into a potentially dangerous TV set with the power on. However, it is quite safe and permissible if you observe reasonable precaution. It is my intention to properly inform you regarding the dangers lurking behind that cover.

The safety interlock in the vast majority of older TV sets consists of the following. The ac line cord is physically tethered to the protective back cover of the set. When the cover is removed, the ac cord is automatically pulled out of a two-pin male plug on the TV chassis. Operating the set with the back removed requires a line cord identical to the one attached to the cover. This is the cheater cord, and it's available from many electronics stores.

In a number of recent TV sets, the cord is not fastened to the back cover. Instead, it can be unplugged from the back of the set before the back is removed. If this is not done first, the cord will automatically disconnect from the chassis when the back cover is taken off. In either case, removing the cover also disconnects the cord, and it takes a deliberate action to plug the cord back into the chassis without the cover. This still ensures that the TV set will not be accidentally exposed and operating; however, in this case, a separate cheater cord is not required.

Today, many TVs with adequate fuse protection have the power cord wired directly into the TV chassis. When working on one of these type chassis, remember to always pull the power cord before touching any component connected to an ac or dc voltage. Exercise extreme care while working around a hot TV chassis.

Basic troubleshooting procedures

Basic voltage measurements

Critical voltage measurements on a transistor, IC, or microprocessor can determine if the component is normal or defective. Voltage measurements on both sides of a coupling capacitor can indicate if the capacitor is leaky. A quick voltage measurement on the main filter capacitor can quickly tell if the power supply source is normal. The supply voltage terminal (V_{cc}) of a suspected IC might indicate a part is leaky or is receiving an insufficient supply voltage.

The most important test for a new transistor is the collector terminal voltage and forward bias measurement. Low collector voltage can indicate either a leaky transistor or improper voltage source (Fig. 5-3). Often, a similar voltage measurement of all terminals indicates that a transistor is leaky. Critical voltage measurements on the emitter terminal can indicate an open transistor or defective emitter resistor with zero voltage.

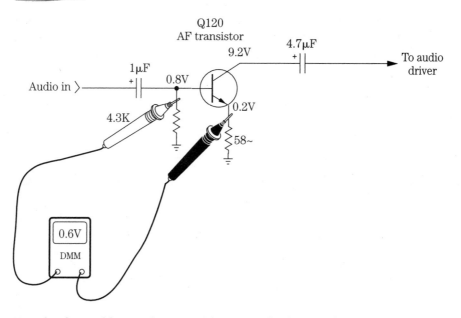

5-3 The forward-bias voltage test between the base and emitter terminals can indicate if the transistor is good or bad.

A forward bias voltage taken between the emitter and base terminals indicates if the transistor is open, leaky, or has a high-resistance junction. The bias voltage of an npn transistor is 0.6 volts, and the pnp germanium transistor is 0.3 volts. No bias voltage between these two elements indicates a transistor is defective. You can quickly check the forward bias on a bunch of transistors in the circuit within minutes. The DMM is the most accurate voltmeter, when the voltage is less than 1 volt at any component.

Basic resistance measurements

Critical resistance measurements are needed when low resistance is in the circuit, of coils, speakers, transformers, relays, or emitter resistors. The resistance measurement across an electrolytic capacitor can indicate a capacitor is leaky or open (Fig. 5-4). The positive terminal of the capacitor should be removed from the circuit for a quick-charge test on any electrolytic capacitor. The capacitor will charge with the ohmmeter leads of a VOM. Slowly, the capacitor will discharge and charge up the other way, if the test leads are left on for a longer time. The smaller the capacity, the lower the charge and discharge. Of course, the DMM numbers will add up and slowly discharge downward, which is very difficult to follow. Use a VOM or FET ohmmeter for capacitor charging tests.

Large resistors have a tendency to change resistance and should be checked with an accurate DMM. Remove one lead for an accurate test. Resistors that operate in high-current circuits might change resistance, overheat, and crack into or show signs of overheating. Especially check power resistors over 10 watts.

The leaky transistor can be identified by resistance measurements from each terminal to the common ground or chassis. A quick low resistance measurement between the collector and emitter terminals indicates that a transistor is leaky. Usually, transistors become leaky between collector and emitter terminals. The resistance

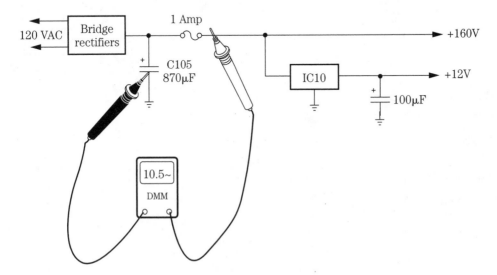

5-4 A quick resistance test across the main filter capacitor can indicate that a capacitor is leaky or that circuits are overloaded.

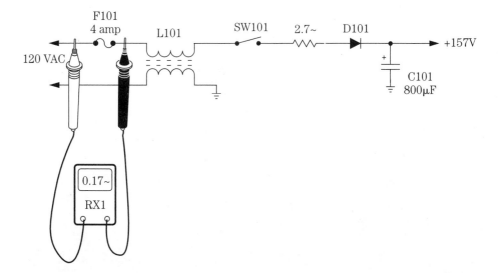

5-5 A quick continuity test across the fuse terminals can indicate if the fuse is open.

reading between emitter and chassis ground might indicate that a resistor is open. Voltage, resistance, and circuit tests upon a suspected transistor in the circuit can determine if the transistor is open or leaky.

Continuity Tests

Actually, a continuity (a state of being continuous) test is a quick resistance test. You can make a quick continuity test to see if a winding is open in a motor, open speaker voice coil, continuity of a coil or transformer and fuse (Fig. 5-5). Often, fuses are tested for continuity, not resistance. The same is true for a lamp or dial bulb.

With a continuity tester, you can test the complete electrical or electronic path, often, just to see if the path is open or closed. The different continuity tests are made with batteries and buzzer, battery and lamp, VOM and DMM low-resistance readings. A quick continuity test can be made with the low-resistance scale of the VOM or DMM. If a resistance of a component is greater than 100 Ω, you should make a resistance test instead of a continuity check.

Basic transistor troubleshooting

Locating and replacing transistors in the radios and TVs might be easy for some people and rather difficult for others. Simple diode and transistor tests in the power supply, horizontal circuits, and audio stages can help you to locate the possible defective part with the digital multimeter. Of course, the video, sync, AGC, vertical, color, and luminance circuits are quite complicated and should be left up to the electronic technician.

For example, if the audio stage is dead in the TV set, you can make several continuity tests with the DMM. Check the continuity of the speaker with a low-resistance measurement (Fig. 5-6). If there is no reading, the speaker cone is open, or cable or

5-6 Check the speaker voice coil continuity with the low-resistance scale of the DMM.

soldered connection is defective. The speaker is probably good with a 2- to 50-Ω reading. The normal 8-Ω speaker will often measure around 7.5 Ω.

If the speaker is normal, check the coupling capacitor between speaker and solid-state circuit by shunting a good electrolytic (100 μF) across the suspected one (Fig. 5-7). The output transistors and audio ICs can be checked by taking voltage, resistance, and transistor tests with the DMM.

Besides taking voltage measurements on the transistor elements, a base bias voltage measurement can indicate if the transistor is normal. The base bias voltage is less than 1 volt and is taken between the emitter and base terminals. The DMM is ideal for this measurement. Silicon transistors have a normal voltage measurement of 0.6 volts, and germanium transistors have a 0.3-volt reading (Fig. 5-8).

If the voltage measurement is zero between these two elements, the transistor is open or leaky. A higher voltage measurement between elements might indicate that the transistor has a high-resistance junction. For more voltage and transistor junction tests, see Chapters 10 and 11 (Fig. 5-9).

Where do you start?

Begin with the symptoms from the picture tube or speaker. No picture, raster, or sound might be caused by the power supply or horizontal circuits. First check for open or blown fuses. Next, measure the voltage across the large filter capacitor (Fig. 5-10).

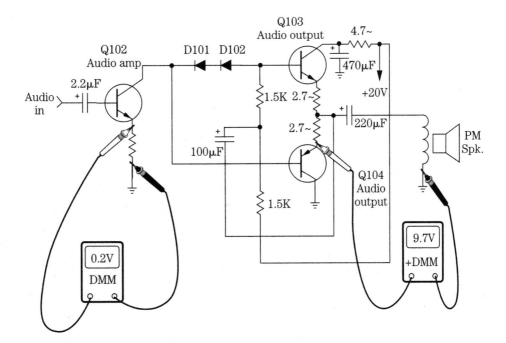

5-7 Emitter and collector voltage measurements on the audio transistors results in a leaky or open transistor.

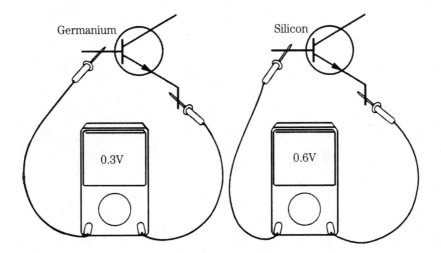

5-8 The forward bias voltage between the emitter and base terminals can help you to identify a npn or pnp transistor.

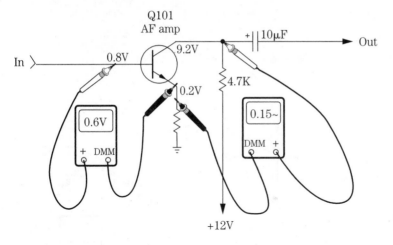

5-9 Test the suspected transistor with voltage and resistance measurements to locate a leaky or open transistor.

5-10 The voltage taken across the main filter capacitor indicates if the low-voltage circuits are functioning.

If the sound is normal with only a thin, bright horizontal line across the raster or screen, there is no or improper vertical sweep (Fig. 5-11). In vertical circuits, it is a little tricky to locate the defective component. But you can take transistor voltage, resistance, and transistor tests on each transistor in the vertical circuits.

The symptom might be no sound or no raster, but the voltage is measured at the low-voltage power supply. Because no sound, raster, or sweep are indicated on the screen, check the horizontal circuits. Measure the voltage applied to the horizontal circuits. Do not check the voltage on the output transistors with any small meters.

Because the horizontal output transistor causes the most trouble in the TV chassis, you can check from the collector (shield) to chassis ground for a leaky output transistor (Fig. 5-12). The resistance with the DMM placed on the diode test should be somewhere between 0.450 to 0.550 Ω. If the resistance is lower, or less than 1 Ω, suspect that the output transistor or damper diode is shorted. Test the transistors in the circuit with the DMM.

If the picture is normal without any sound, go directly to the audio section. Most audio problems are developed in the audio output stages. Check the speaker continuity and output transistor tests. The sound stage might be normal in a no raster/no picture/no audio situation because, in most chassis, the horizontal circuits must function before the audio stages can work.

Always, start to check the TV chassis by looking for a trouble symptom on the screen or by the speaker. Feel components to see if they're overheating, such as transistors, resistors, capacitors and ICs. Smelling overheated or burned components might turn up a defective part. Don't forget to use the DMM. It's best to leave high-voltage and picture tube circuits alone, unless you have the know-how or test equipment to make these tests.

Various symptoms

Most TV symptoms can be found by looking at the TV screen and listening to the speaker. Horizontal TV problems might result from no raster, no high voltage,

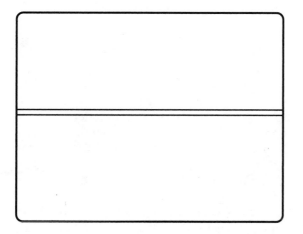

5-11 A white horizontal thin line indicates that there is no vertical sweep.

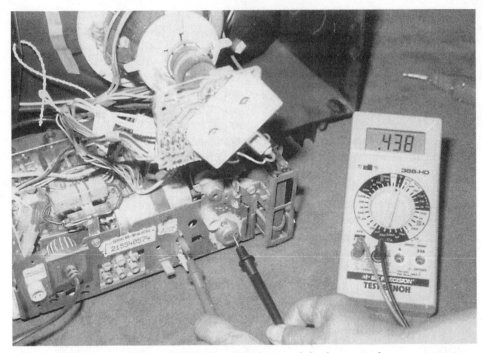

5-12 A transistor or diode test from the metal case of the horizontal output transistor to the common ground indicates that the transistor or damper diode is leaky.

intermittent chassis, shutdown, and no picture. The raster has black areas on each side of the picture, indicating improper width, that can be caused by low-voltage source, improper HV at the CRT, and defective horizontal circuits (Fig. 5-13). A white vertical line indicates an open yoke or no horizontal sweep. Of course, high voltage is present with the vertical white line. Sometimes a dark screen with high voltage indicates video problems, not horizontal, especially with normal sound or audio.

A white horizontal bright line results from no vertical sweep. Improper vertical sweep or only four inches of picture indicates insufficient vertical sweep. Improper vertical sweep can be caused from a defective output transistor, IC, electrolytic capacitor; a change in resistance; or an improper voltage applied to the vertical circuits (Fig. 5-14).

The no sound/no raster/no picture symptom can be caused by a defective low-voltage source. Check the dc voltage across the main electrolytic filter capacitor. If there is no voltage, look for a leaky electrolytic capacitor, shorted silicon diodes, open isolation resistor, or an open line fuse. Check the continuity of the fuse with the DMM. When both line and B+ fuses are blown, suspect a leaky horizontal output transistor, damper diode, and flyback.

With a fuse-keeps-blowing symptom, suspect that the silicon diodes, electrolytic capacitor, or horizontal output transistor or flyback are leaky. Clip a 100-watt bulb across the open line fuse (Fig. 5-15). Remove the horizontal output transistor and notice if the light goes out. If so, check the horizontal output transistor, damper diode, and flyback. Suspect that a transformer or IC voltage regulator is leaky when, after removing the transistor, the light stays on.

A normal picture with distorted or no sound indicates defective sound output circuits. Check for leaky or open transistors if the audio is weak and distorted. For weak sound, check the coupling capacitor, transistors, and ICs. Do not overlook the possibility of a dried-up coupling capacitor between the speaker and output transistor or IC if the sound is weak. By simply checking the symptoms on the TV screen and speaker, you can isolate what section the defective component lies in.

Removing the chassis

Before attempting to remove the chassis in a console TV, disconnect all leads and cables that connect to the chassis. This can include speaker cables, degaussing coil, antenna cable, high voltage cable, and deflection yoke plug or wires. If in doubt where they connect, make a small sketch and locate the various connections that are unplugged from the chassis. Be careful not to damage or move

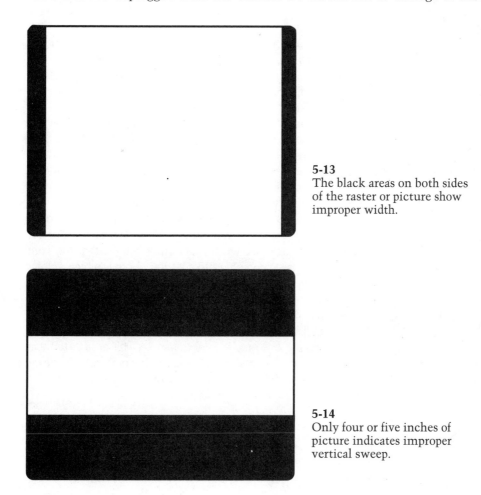

5-13
The black areas on both sides of the raster or picture show improper width.

5-14
Only four or five inches of picture indicates improper vertical sweep.

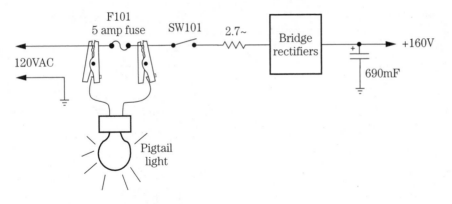

5-15 Check the ac line fuse with a pigtail light. When it goes out, the overloaded component has been disconnected.

components on the chassis. Be very careful not to drag the chassis across the wooden shelf to prevent damaging surface-mounted parts mounted underneath on the PC wiring.

In today's portable TVs, you can loosen up a couple of side screws on each side of the chassis and slide the chassis back, so you can see the components on top and bottom sides of the chassis. Often, when the back cover is removed from a portable TV, the PC board chassis can readily be seen and serviced. Simply tip the TV on its side and take critical tests (Fig. 5-16).

The TV chassis in a console model can be released by removing the side or bottom bolts and metal screws. You can service the chassis in the cabinet by placing the whole cabinet on a TV service cart. Simply pull the chassis up enough to get at the bottom PC board. Sometimes the chassis can be turned over with cables attached, to take voltage measurements in a certain section of the chassis (Fig. 5-17).

Most service data clearly identifies what connectors exist between the chassis and the speaker and picture tube. Usually, two separate wires run from the chassis to the speaker—each with a quick-disconnect device. Sometimes, these two wires interconnect through a two-pin plug. The picture tube usually has a multipin plug for the same purpose, and in some color TV sets, two such plugs. In addition, there is a high-voltage lead going to a snap button on the picture tube glass. This also must be disconnected, preferably a few minutes after the set has been switched off to allow time for the residual high voltage to dissipate. Otherwise, an unpleasant jolt will be in store for the bare hand touching this lead. Incidentally, a large number of TV sets have a quick-disconnect on this lead at the chassis end, making it so much easier to handle. This applies to those receivers in which the picture tube is mounted to the cabinet. In some sets, however, especially some of the portables, the picture tube is mounted on the chassis, and no disconnecting of any picture tube wires or plugs is required.

Testing by substitution

Sometimes you might have to substitute parts to find the defective one. Tubes were subbed in the tube chassis to determine if the old tube was normal or defective. Electrolytic capacitors can be subbed or clipped into the circuit to eliminate hum in

5-16 Simply removing the back cover of a portable allows you to see most of the electronic components.

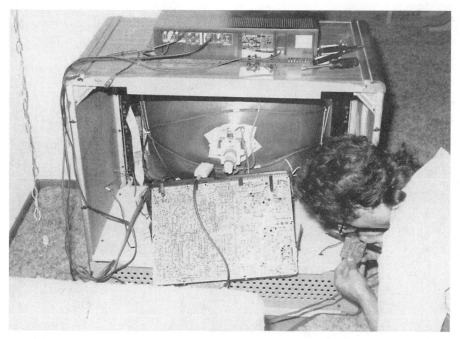

5-17 The chassis within the console TV can be removed, tilted upward for voltage, resistance, or component replacement.

the audio or picture and raise the voltage source. When replacing electrolytic capacitors and you do not have the same value, sub one with a higher capacitance and working voltage.

For instance, if a 470-µF 25-volt electrolytic capacitor is suspected, you can clip a 1000-µF 50-volt capacitor across it. In fact, you can replace the 470-µF capacitor with a 1000-µF electrolytic without any difficulty, provided that room is available (Fig. 5-18). Always remember to replace a defective electrolytic capacitor with one that has the same (or higher) values (capacitance and working voltage). Bypass and coupling capacitors can be replaced in the same manner.

Appropriate "universal" transistors and ICs can be substituted. Simply look up the original part number in a RCA, GE, NTE, or Sylvania universal semiconductor replacement manual. Always use exact replacements for critical or safety-marked components.

Locating components on the chassis

After isolating the various symptoms with the help of the screen or speaker, locate the correct circuit on the TV chassis. Locate the low-voltage power supply, which contains a bridge rectifier, silicon diodes, and a large filter capacitor. Locate transistors mounted on separate heatsinks, close together for vertical output circuits. Likewise, check for vertical output IC circuits on a separate heatsink. Locate the horizontal transistor on a heatsink or directly mounted on the metal chassis, near the flyback or horizontal output transformer.

Main electrolytic filter capacitor

5-18 Large electrolytic filter capacitors can be replaced with parts that have higher capacitance and working voltage, if original parts are not available.

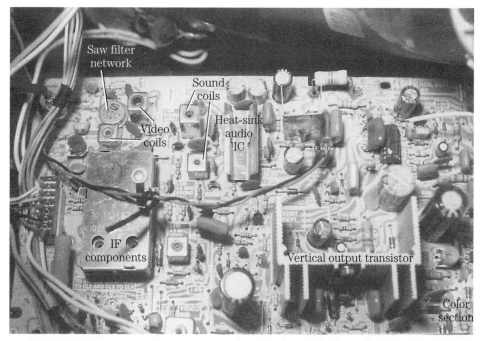

5-19 Small components, shielded parts, and heatsinks can help you to locate the various TV sections.

The TV input circuits can easily be located by locating the varactor tuner and shielded IF cables. A large IC might include several different circuit functions and can be traced to the IF/SIF/sync/AGC/video circuits nearby. Small shielded transformers (IF and video) and the saw filter network follow the shielded IF circuits (Fig. 5-19).

Check for audio shielded takeoff and discriminator coils (with a IC heatsink) nearby for the audio circuits. Some audio output ICs are mounted on a heatsink, and others have a heatsink glued to the top of the IC with metal cooling fins. If in doubt, trace the audio circuits from the speaker terminals back to the corresponding IC or transistor output circuits.

Common-sense servicing

Many defects and defective components can be located with sight, small sounds, and safely touching the PC board or components. Wiring connections might produce a burning smell or smoke rising from the various components. Water or a soft drink spilled down inside the portable TV cover might cause the PC wiring to burn. Cutting out a portion of the board and cleaning the chassis might prevent further breakdown.

You might hear arcing around the picture tube anode or high-voltage tripler unit, causing the fuse to open. Besides checking the speaker sounds, you might hear the high voltage rise with the expansion of the yoke winding, indicating that high voltage is present without a raster on the screen. A high-frequency noise from the flyback transformer might indicate the horizontal oscillator frequency is way off and needs adjustment or repair.

A close inspection of a cracked or burned resistor might help you to locate a corresponding defective component. After replacing a defective tube or transistor, the circuits might still be dead because these components might cause voltage-dropping bias resistors to overheat. A filter capacitor with a white substance oozing from the pin connectors is defective. Burns around the flyback transformer might indicate high-voltage arcing (Fig. 5-20).

A large-wattage resistor that is too hot to touch might indicate a shorted capacitor. A power horizontal output transistor that feels too hot might be overloaded with a defective horizontal output transformer. A warm tripler unit might indicate excessive current. Remember, besides test equipment, common-sense servicing begins with sight, smell, and sound.

Heat and coolant

The electronic technician might use heat or coolant to make a solid-state component act up in the TV chassis. Heat can be applied to the TV chassis with a blow dryer. When the transistors or ICs reach a certain degree of heat, they will break down. Heating/cooling is most effective with intermittents (Fig. 5-21).

Heat can be applied to a particular component with a light bulb, soldering iron, hair dryer, or heat blower. Sometimes by holding the soldering iron or gun close to a capacitor or transistor, the extra heat causes it to act up. After several applications of heat with the hair dryer or blower, the capacitor, transistor, or IC show its defective side.

5-20 Burned or cracked areas of the flyback can indicate that a transformer is leaky or arcing.

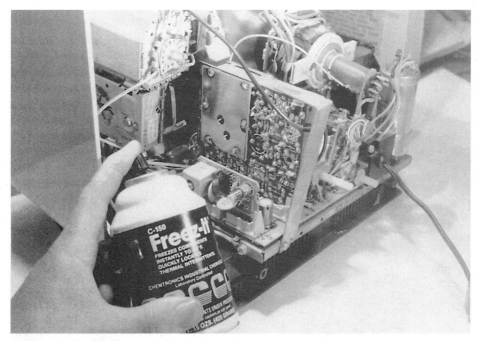

5-21 Cans of coolant and a hair dryer can help you to locate intermittent components.

Likewise, applying several coats of coolant on the component might make it act up. Because transistors, capacitors, resistors, and ICs consist of solid material, a spray of coolant might make the component go into the intermittent or normal state. Cold spray or coolant comes in a can like hair spray or paint, with a plastic nozzle applicator.

You might find that applying a coat of coolant, then blowing heat on the suspected component will solve the most difficult intermittent condition. Be careful when applying heat to not melt components. Apply coolant right on the suspected part. Sometimes you can locate cracked PC wiring or a poorly soldered connection with several coats of coolant.

Intermittent boards and poorly soldered connections can be located with the soldering iron. Try to locate the area where the intermittent occurs by taking voltage measurements or pushing up and down on the PC board. Then, solder all connections in that area (Fig. 5-22). Be careful not to spill solder over the board or solder together two junctions. Check the soldered junctions with a magnifying glass.

Replacing universal transistors

Like all parts, transistors should be replaced with originals. When an original type is not available, most transistors can be replaced with a universal transistor, without any problems. Look up the part number of the transistor in the universal semiconductor-replacement manual of RCA, GE, Sylvania, Workman, and NTE (Fig. 5-23). Sometimes the transistor has both numbers and letters stamped on the body area. Usually, the part number is listed in the manufacturer's service manual or Howard Sams Photofacts.

5-22 Sometimes touching up the IC terminals with a battery soldering iron can solve intermittent problems.

5-23 Look up the semiconductor part number in a universal semiconductor manual for part replacement.

For instance, you might find a 5-V regulator transistor leaky from the emitter to collector terminal in a JC Penny 685-2520-00 chassis. The number found on the transistor was D637. By checking the RCA universal replacement manual, you know that the regulator transistor is used in medium-power circuits in TVs, CBs, and VCRs. The maximum collector voltage lists a universal SK3911, NTE16, or ECG16 transistor for replacement. Because 13.9 volts was applied to the collector terminal, the universal replacement (SK3911) worked fine in the voltage-regulator circuit.

Replacing universal ICs

Replacing ICs is very similar to replacing transistors. Some ICs are not listed in the universal replacement manuals and must be obtained from the manufacturer or service depot. Always replace ICs with components that have the original part number, if they are available.

Look at the top of the IC for a part number. If a schematic is handy, look it up. In a Sony AVM255 portable TV, a malfunctioning dual stereo channel output IC resulted in weak and distorted audio. The supply voltage was low, indicating that the IC might be leaky. It was. The part number STK436TV was checked in several different universal replacement manuals, but no replacement (Fig. 5-24). Howard Sams Photofacts did not list a universal replacement. This meant the original part must be obtained from the manufacturer or mail-order firm. The STK436TV IC was purchased from MCM Electronics in Centerville, Ohio for $10.90. The part was delivered within three days.

Obtaining the correct schematic

If you service or repair of your own TV set, you should have a schematic diagram. This service literature can be secured from the manufacturer. Of course, this literature is not free. Simply write to the manufacturer and give the model number of your TV, ask where the schematic can be purchased, and for what cost (Fig. 5-25). You might have a copy made from a schematic located at your local TV repair shop.

Most electronic technicians that service TVs can obtain service literature from the manufacturer, service depot, or wholesale outlets. Service literature is usually furnished with a small fee or sometimes at no cost to the TV warranty station. The electronic technician who does warranty service for several different manufacturers, also receives information concerning the latest problems and cures to keep that TV operating.

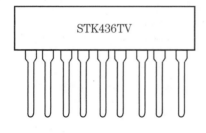

5-24
Look on the body of IC for the part number. Here is a dual stereo audio output IC (STK436TV) in a Sony TV chassis.

5-25 If you plan to do a lot of TV servicing, the schematic diagram is a must-have item.

Howard Sams PhotoFacts

Many TV technicians subscribe to the Howard Sams PHOTOFACTS, which supply many of the latest schematics for most TV manufacturers. A yearly index is provided, which lists all of the TVs that PhotoFacts provides service diagrams and information. Besides providing a list of TV schematics, authorized PhotoFact distributors, manufacturers, and test-equipment manufacturers are listed in the index copy (Fig. 5-26).

Schematic diagrams of the TV circuits, tuner, display, low-voltage power supply, interconnecting diagrams, remote transmitters, on-screen display, and picture tube circuits are included separately in the PhotoFacts. Besides schematics, the Photo-Facts contain actual photos of parts layouts, resistors, capacitors, diodes, transistors, and ICs with a location guide. Also included are complete lists of semiconductors, capacitors, controls, power and special resistors, coils and transformers, cabinet and cabinet parts, and miscellaneous components. Often, the manufacturers part numbers are listed with item symbols and replacement data.

Howard Sams PhotoFacts can be purchased directly from the company at 2647 Waterfront Parkway, East Dr., Indianapolis, IN 46214. Also, many electronic distributors and mail-order firms have the Sams PhotoFacts in stock or they can get them for you. Schematics are a must if you are planning any type of electronic servicing.

Typical solid-state troubleshooting charts

Tables 5-1 and 5-2 show two different troubleshooting charts. Remember to leave the most difficult repairs to the electronic technician. If in doubt, don't touch it.

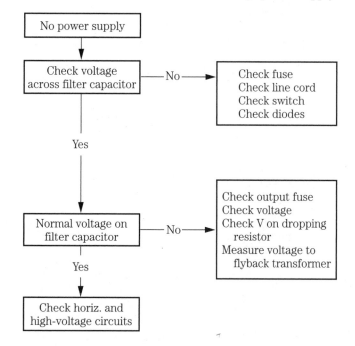

5-26 Howard Sams PhotoFacts provide valuable service information, schematics, and photo data for various TVs.

Table 5-1. Flowchart of a TV low-voltage power supply.

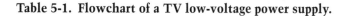

```
            ┌─────────────────────┐
            │   No power supply   │
            └─────────────────────┘
                      │
                      ▼
        ┌─────────────────────┐          ┌──────────────────────┐
        │   Check voltage     │   No     │   Check fuse         │
        │ across filter       │─────────▶│   Check line cord    │
        │    capacitor        │          │   Check switch       │
        └─────────────────────┘          │   Check diodes       │
                      │                  └──────────────────────┘
                     Yes
                      │
                      ▼
                                          ┌──────────────────────────┐
                                          │ Check output fuse        │
        ┌─────────────────────┐   No      │ Check voltage            │
        │ Normal voltage on   │──────────▶│ Check V on dropping      │
        │   filter capacitor  │           │    resistor              │
        └─────────────────────┘           │ Measure voltage to       │
                      │                    │    flyback transformer   │
                     Yes                   └──────────────────────────┘
                      │
                      ▼
        ┌─────────────────────┐
        │ Check horiz. and    │
        │ high-voltage circuits│
        └─────────────────────┘
```

Table 5-2. A flowchart of the various symptoms, circuits, possible causes and replacements in the TV.

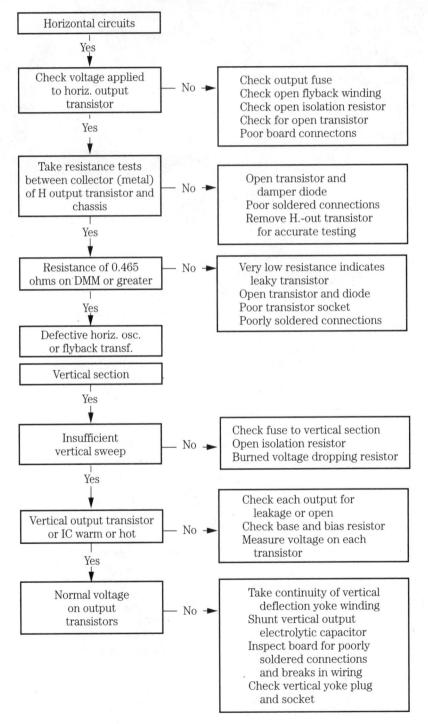

Table 5-2. continued.

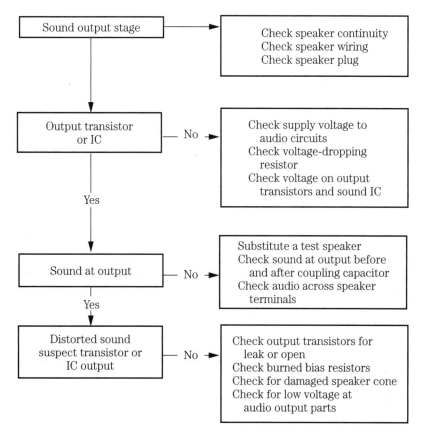

6
CHAPTER

Operating adjustments

The next logical step in TV troubleshooting is a knowledge of the operating controls—not only the knobs on the front of the receiver, but those mysterious-looking adjustments in the back.

Front controls

The front operating controls typically consist of contrast, brightness, tint, color, sharpness, bass, treble, and balance (Fig. 6-1). The latest 27-inch console TV also has power, volume up and down, channel up and channel down, video, audio, and set-up controls in the front-panel area. Today, large-screen TVs have many of these set-up controls on a menu that can be adjusted with the remote control. The rest of the set-up adjustments are at the rear of the chassis.

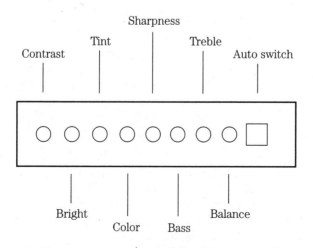

6-1 The front panel controls can be enclosed behind a lid or out in plain view for TV adjustments.

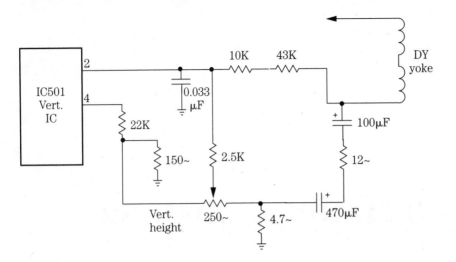

6-2 The vertical size or height adjustment is adjusted to fill the screen with $\frac{1}{2}$ inch over the scan.

Vertical size

The typical TV chassis has a vertical size or height control, to fill out the screen completely with room to spare. In the back of the set or PC board, this adjustment is called the *vertical size control* (Fig. 6-2). The vertical size control can be adjusted to fill out the picture, top and bottom. Connect a dot-bar generator to make this adjustment with a cross-hatch pattern or regular TV test pattern. Set the contrast and brightness controls to maximum. Adjust the vertical size control so that the large circle reaches the edge or $\frac{1}{2}$ inch over frame and cross-hatch lines evenly spaced.

Another important consideration before adjusting the vertical size is the avoidance of distortion. The normal picture format (or aspect ratio) by technical and NTSC standard is four units wide by three units high; that is, for every four inches of picture width is a corresponding three inches of picture height. Under these conditions, a circular emblem will look like a perfect circle, not egg shaped. When adjusting the vertical size, be careful not to stretch the picture to the point of distortion. This is best done with some circular object on the screen. Some stations still show a circular test pattern a few minutes before going on the air; however, it is not necessary to wait for a test pattern—many emblems and other commercial symbols shown between programs are circular and will serve for vertical size adjustments. Possible alternatives for height adjustments, therefore, are as follows:

- If the picture height is slightly below normal, adjust the vertical size control until the picture just covers the screen height. Check for nondistortion by observing the circularity of a suitable object on the screen, preferably a large one so that it's easier to make correct adjustments. For all adjustments where observation of the screen is necessary, a mirror is essential. The professional TV servicer has a special mirror on a stand for the purpose, but any mirror propped up in front of the screen so that it can be viewed from behind the set is quite satisfactory.

- If the control setting for full-screen height is at or near its extreme position (little rotation remaining), the vertical amplifier stage is not operating properly.

Vertical linearity adjustments with crosshatch lines

After setting the height control, check the screen or picture for poor linearity. The picture might be squashed at the top and long at the bottom, or vice versa. Connect a color bar generator to the antenna terminals. Tune in channel 3 and switch to a crosshatch pattern. Notice if the lines are all evenly spaced. If not, adjust the vertical linearity control so that all spaces are equal across the screen. Sometimes the height and linearity controls must be adjusted together to require correct vertical linearity and height.

Vertical size adjustment

1. Switch the TV on.
2. With a mirror positioned so that you can see the picture from your position behind the set, reduce the V-size until the picture is too small to cover the screen, both at the top and bottom; that is, until some blank screen shows.
3. Wait for a circular pattern or emblem to appear on the screen and observe its symmetry. It is quite easy to see whether or not a round shape is circular. Figure 6-3 is an example of patterns you might encounter during this adjustment.
4. If the pattern resembles Fig. 6-3A (the two halves symmetrical), adjust the V-size to increase the pattern or display until all four quarters are as identical as possible, or until the circle is no longer flattened at the top and bottom. This will produce the closest approach to a perfect circle. (Some minor irregularity, because of the horizontal circuits and other imperfections, might prevent the attainment of a perfect circle, but if this deviation is very slight, no further correction is necessary.) The pattern should look pretty much like that in Fig. 6-3D.
5. If the final, most nearly circular pattern is obtained before the screen is fully covered in the vertical direction or, conversely, if you have to turn the V-size control to the point where the picture runs beyond the top and bottom edges of the screen to get a circle, then the width (horizontal size) adjustment is incorrect and will have to be corrected before completion of

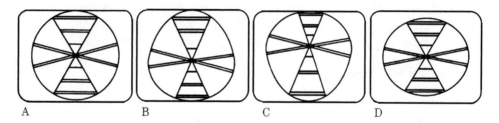

A   B   C   D

6-3 The figure shows the correct test pattern adjustment (A), squashed at the bottom (B), long at the bottom (C), and flat on both top and bottom (D).

the vertical adjustment. However, at the moment, I am proceeding on the assumption that only the vertical circuits require adjustment.

Vertical linearity

If the reduced image on the screen resembles Fig. 6-3B or 6-3C, a second, closely related vertical adjustment must also be made. This control varies the vertical picture symmetry, usually called *vertical linearity*. Although it is difficult to tell a nearly perfect shape by viewing an action scene on the screen, such is not the case for distorted proportion. For example, Fig. 6-3B corresponds to excessive stretching of the top of the picture; people seem to have very short legs and their heads seem to come to a point. The opposite extreme (Fig. 6-3C) produces long-legged, high-waisted people with very low foreheads. Correction of either requires the adjustment of the V-lin (vertical linearity) control, again using a circular pattern of some kind, plus the aid of a ruler and a grease pencil, as follows:

1. Mark a horizontal line across the screen midway between the top and bottom; use a ruler for a fairly accurate division of the height.
2. Adjust the V-size control, as previously, for a circle smaller than the full-screen height, exposing blank space at the top and bottom.
3. Adjust the V-lin control until the upper and lower portions (above and below the painted line) are as nearly of the same height as you can see (measurement is not necessary).

In recently manufactured TVs, the vertical and horizontal hold controls are eliminated. The new vertical deflection system consists of a countdown-divider system for generating the vertical signal. The divider system has an internal frequency-doubling circuit that doubles the horizontal oscillator signal within the countdown divider. Twice the horizontal frequency or two clock pulses are supplied to the vertical countdown-divider system during one horizontal scan line period. Because of the divider system, no vertical frequency adjustment is needed. The vertical sync and frequency is controlled inside the vertical and horizontal countdown IC.

Vertical roll

In a normally functioning TV receiver, the picture will at times slide up or down one frame. This might happen when the set is first switched on, when channels are changed or when programs are switched at the transmitter. At all other times, the picture actually does not "want" to slip or roll because it is controlled by and synchronized with a signal from the transmitter (vertical sync, see Chapter 2). If, however, a TV set exhibits frequent rolling, requiring repeated adjustment of the V-hold control, the fault is most likely in the sync circuits. Replacing the large IC that contains the sync circuits in the latest TV chassis cures most sync problems.

Vertical foldover

Vertical foldover often occurs at either the top or bottom of the picture. Actually, part of the picture is folded over or rolled up, and might not fill out the screen. Vertical foldover problems occur in the vertical output and feedback circuits. Check the vertical output transistor, IC, and yoke coupling capacitor. Check each base and emitter bias resistors when the output transistors are out of the circuit.

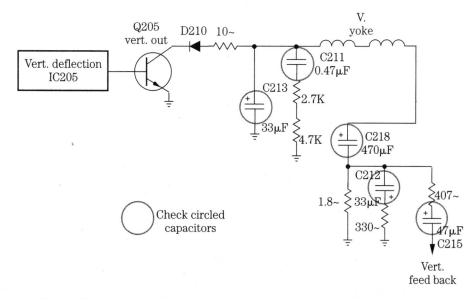

6-4 Vertical foldover can be caused by a defective vertical output transistor or IC (C211, C212, C213, and C215, in this case).

You can find a defective component in the vertical feedback circuits causing vertical foldover. Check for leaky or dried-up electrolytic capacitors and associated components. Check capacitors C211, C212, C213, C215, and C218 in the vertical bias and feedback circuits for possible vertical foldover (Fig. 6-4).

Vertical crawling

Vertical crawling exists when lines bunch together and slowly crawl up the raster or picture. Check for defective filter capacitors in the low-voltage power supply, decoupling capacitors, or a defective low-voltage regulator transistor or IC.

Notice if the raster has a dark band going up the screen. Turn off the chassis and shunt each large filter capacitor in the low-voltage circuits. Clip the new capacitor across the suspected one. Check all capacitors in the power supply in same manner. Do not overlook a defective electrolytic capacitor from vertical output to the yoke winding. Always shut down the chassis so as not to damage semiconductors in other circuits. Charging and discharging electrolytic capacitors might shoot up the voltage and damage components in corresponding voltage sources.

TV adjustments

Some of the early color TV chassis have the vertical, horizontal, contrast, brightness, color, and tint controls inside a hinged lid. Gradually, these controls moved to the rear of the chassis in portable TVs (Fig. 6-5). Today, in the solid-state color chassis, both the vertical and horizontal controls are not located in voltage-controlled oscillators and fixed reference voltages. In fact, with the remote-control TV, you might find that all functions are controlled by the system-control circuits.

Vertical   Horizontal   Vertical
hold   hold   preset

6-5 The horizontal, brightness, and contrast controls are at the rear of the black-and-white TV chassis.

Black-and-white adjustments

The black-and-white portable TV adjustments are located in the rear apron. You might not find a width or horizontal hold adjustment. The brightness and contrast knobs stick out the back with a vertical hold control with a preset adjustment (Fig. 6-6). In early transistor TVs, the vertical height and linearity controls were just inside the cover on the rear apron. The horizontal hold control turned a slug inside the horizontal oscillator control for horizontal lock-in (Fig. 6-7).

In newer black-and-white TVs, you might find separate vertical oscillator, driver, and output transistor stages or ICs. The vertical oscillator, drive, and output circuits might be contained in one IC or combined with other circuits in one large IC (Fig. 6-8). Usually, the vertical IC has a long bar-type heatsink. Again, the vertical height and linearity controls are on the main chassis with the vertical and horizontal hold at the rear of the chassis.

Color solid-state vertical adjustments

In yesterday's color chassis, the vertical circuits had vertical height, linearity, and hold controls. The vertical hold control was located at the rear of the chassis with the height and linearity having preset screwdriver-type controls on the chassis. Only the vertical hold control stuck out the back of the chassis (Fig. 6-9). The vertical oscillator and vertical amp were found in the deflection IC, which also contained the vertical sync, regulator, AFC detector, horizontal oscillator, flip-flop, sync gate, and X-ray

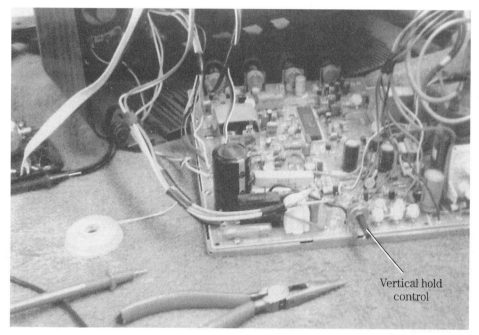

6-6 The vertical hold control is located at the back of the earlier TV chassis.

6-7 The horizontal slug-tuned coil is the horizontal control in the black-and-white TV chassis.

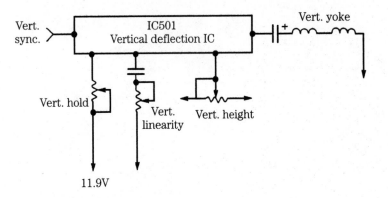

6-8 Today, the vertical deflection IC might contain all of the vertical circuits.

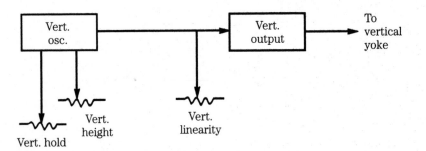

6-9 A block diagram of a vertical oscillator and output transistor with vertical hold, height, and linearity controls.

protector (Fig. 6-10). In later vertical color circuits, the vertical centering, size, and hold controls operated from a sweep oscillator IC. Only the vertical hold control was located on the rear apron. In today's color TV chassis, the vertical and horizontal controls are unnecessary because of a countdown synchronization locked in circuits.

Sync failure

Today, the semiconductor sync circuits are taken off at the video circuits and sync signals are sent back to input terminal 27 of the IC—video/sync/AGC/color/APC/vertical and horizontal deflection IC501 (Fig. 6-11).

Inside IC501, a sync separator circuit separates the vertical and horizontal signals. The vertical output signal (pin 26) is taken from pin 25 of the input terminal of the vertical circuits. A horizontal sync signal from the sync separator is connected internally to the horizontal oscillator circuit. Check for correct voltages on the IC and replace IC501 if poor horizontal or vertical sync is noted.

Of course, in such a case, the picture quality will also suffer, most often exhibiting a lack of highlights and contrast. Although advancing the contrast control might result in some improvement to the picture, this adjustment might actually be detrimental to the normal sync output; thus, chronic picture rolling in a set having a weak picture

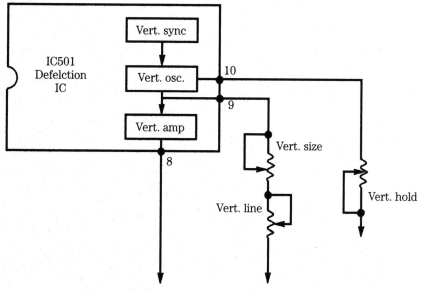

6-10 A block diagram of a vertical deflection IC with proper vertical circuits.

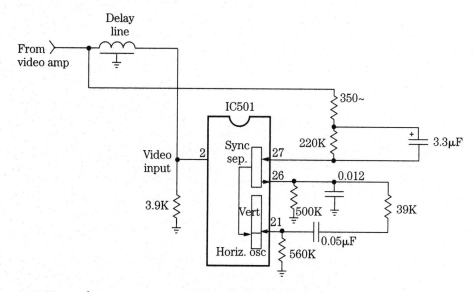

6-11 Vertical sync circuits in a recently manufactured TV.

might be caused by a subnormal video amplifier. Also, although a poor picture might result from anything from the antenna to the video amplifier, it is rather easy to tell which portion of the set is responsible. A weak, "thin" picture that is otherwise clean is almost certain to be caused by a poor video amplifier, while substandard functioning in the TV tuner or in the IF amplifier is almost certain to produce a snowy, perhaps wavy, picture.

"Freak" sync trouble

There is one other potential case of sync problems worth mentioning here—interference from another TV station on the same channel. Standard FCC channel allocations are such that under normal conditions, the geographic separation between the stations operating on the same channel preclude any likelihood of the distant station interfering with a local station on the same channel. Modern TV sets have, in addition, a certain amount of built-in immunity from this possible interference because they tend to favor the much-stronger local station; however, during some not-too-frequent freakish atmosphere conditions, when distant signals tend to come in rather strongly, interference might result. This is because although technical standards (frequencies, sync, sweeps, etc.) are the same throughout the United States, very minor and quite tolerable differences between those standards will cause severe interference with the local station sync, consequently, with the picture as a whole; however, because these are very transient conditions, occurring on some stations only, nothing should be done about it. It is described here merely to acquaint you with the phenomenon, and caution you not to rush into making adjustments.

AGC adjustments

You might find the AGC or AGC delay adjustment control in some of the early solid-state TVs. Simply tune in the strongest or local TV station in your area. Push in the AFT and Auto Color Switch. Turn the AGC delay control fully clockwise, then turn it counter clockwise until the noise just disappears from the scene. If the AGC control is left too high, the strong station might overload the picture. Some TV manufacturers might have a certain voltage setting for AGC delay- line adjustment.

Horizontal size or width

If the picture or raster is dark on both sides, improper width might result from a defective horizontal output circuit, flyback, pincushion circuits and low-voltage power source from the power supply (Fig. 6-12). Some TV chassis might have a width control to help enlarge horizontal sweep. Too low of a voltage at the CRT can cause improper width.

An improper adjustment of the B+ or high voltage can cause the sides to pull in. Check the HV at the picture tube and the low voltage applied to the horizontal output transformer or flyback. Check for a leaky regulator transistor in the low-voltage power source when the voltage cannot be raised or lowered. Test the transistor regulator and zener diode in the B+ circuits. Do not overlook the possibility that the primary winding of the horizontal driver transformer is shorted.

Potentiometer (pot)   This volume-control adjustment might be a knurled, round shaft, like that inside the knobs of the ordinary radio, or it might be a slotted shaft designed for screwdriver adjustment. In either case, it is an adjustable pot that can be ro-

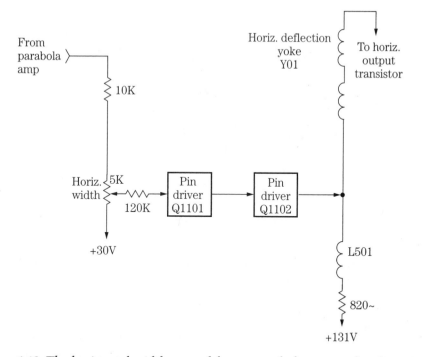

6-12 The horizontal width control from a parabola amp to the pin-cushion circuits.

tated less than one full turn. It is the simplest to adjust and can easily be reset to its original (before adjustment) position. As mentioned, it is wise for the beginner to always observe and, where possible, record the position of an adjustment before changing it. You can always go back to the original setting if you feel that it was the best.

Slowly rotate the shaft (using a mirror to watch the picture) until the width is less than the full screen. If the picture extends more or less equally on both left and right edges, rotate the shaft in the opposite direction until the picture just barely extends beyond the edges of the screen. About ¼- to ½-inch is all that should be wrapped on each side. The correct width setting is that which does not mask any part of a line of printed matter (the list of actors, etc. displayed at the beginning of a movie is most suitable).

Slug A slug-type width adjustment is used in many TVs. It requires a different technique, as well as some precautionary measures. The word *slug* refers to a carbon-like rod that can be made to slide in or out of fiber sleeve on which a spool of wire is wound. The whole assembly is called a *width coil* or a *slug-tuned coil*. The tuning or adjustment of width is made by inserting the blade of a special tool (or midget screwdriver) into the protruding slot of a threaded screw-like shaft. Figure 6-13 is a simple sketch of such a coil.

In contrast to the previous pot adjustment, the slug can be turned as many as eight or more times. In fact, there is no stop on this adjustment, hence the precaution: do not turn a slug more than three or four turns in either direction. To do

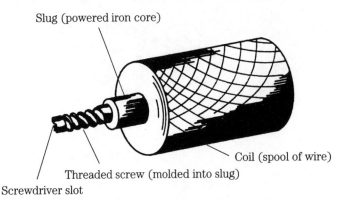

Slug (powered iron core)

Coil (spool of wire)

Threaded screw (molded into slug)

Screwdriver slot

6-13 A typical slug-tuned type of coil in the earlier TV chassis.

so might unscrew the slug completely, letting it fall out or, worse, fall into the dark recess of the TV chassis. Whichever way it falls, the job of replacing it is messy, certainly one to be avoided by the beginner. Fortunately, TV design is such that this slug is normally in an approximately midposition, allowing a minimum travel of five or six turns in or out—certainly enough for any width adjustment in the field. Normally, most width coils are designed so that turning the slug out (counter-clockwise from the screwdriver end) will increase the width, and vice versa. The procedure, therefore, is the same as the pot shaft rotation, except that a few full turns might be required.

The correct procedure for adjustment is: Rotate the slug clockwise until both left and right edges of the picture are visible; the picture is now too narrow. Rotate the slug in the other direction until a slight overlap or wrap is achieved on each side. Of course, too much overlap cuts off the picture.

Points   In this method of width adjustment, a movable voltage lead can be attached to various voltage points. By changing the voltage applied to a portion of the horizontal amplifier, the width of the picture can be varied slightly. Usually, there are three or four tie points to select. In some cases, not connecting the movable jumper lead to any voltage point provides a possible setting for the horizontal width. Because the movable jumper is usually insulated, this procedure can be made with the TV under power.

Other width adjustments   In some TVs, a mechanical (magnetic) adjustment is provided for picture width control. No two of these are alike, and the adjustment should not be attempted without some sort of service instructions. Two of the most common mechanical adjustments are the width sleeve and the width tabs.

The width sleeve is just that—a metallic sleeve on the rear of the neck of the picture tube. This sleeve can easily be adjusted with a combined sliding-rotation motion (it might sometimes stick to the glass until dislodged) until the correct width adjustment is obtained. There is no danger from either heat or high voltage in this procedure, but neither is there room for carelessness. The sleeve sliding along the neck of the picture tube also affects horizontal linearity.

Width tabs function in much the same manner as the sleeve. The tabs have the same effect on width as a sliding sleeve. Because there can be a great variety of mechanical means of varying picture width, no standard, uniform procedure can be given, except as described.

Horizontal linearity   As mentioned previously in connection with the procedure for horizontal width adjustment and correction, the horizontal linearity adjustment is most conveniently made at the same time. In some sets, the two interact so that it is almost mandatory to do them together, unless a nonlinearity problem exists without any accompanying reduction in width. In that case, it might be necessary to reduce the width to be able to see the nonlinearity.

Horizontal linearity can best be understood by referring to the sections on vertical linearity. In simplest terms, horizontal linearity means that the left and the right halves of the picture are symmetrical. More specifically, and this applies to the vast majority of TVs, *horizontal linearity* means that a circle does not appear like an egg, and that the right side is neither stretched nor compressed. This is not always obvious when the picture is wrapped around the tube sides; therefore, nonlinearity is best shown when the width is reduced to a little less than the edges of the screen. As in the case of the vertical linearity, a circular object is most helpful in observing horizontal linearity, although this is not a must. It should also be kept in mind that although the standard aspect ratio (picture width versus height) is four to three—that is, the picture is a rectangle, not a square— a circle will appear (not an egg lying down) on a properly adjusted TV set.

Horizontal linearity adjustments are not the same on all TV sets; however, most sets have one of three types of adjustments. In earlier TVs, a potentiometer adjustment on the rear apron of the set provides correction for nonlinearity. In other sets, a slug adjustment, identical in appearance with the width slug adjustment described earlier, is used. In still other TVs, a metallic sleeve on the tube neck, by itself or in conjunction with a shaft adjustment, serves the same purpose.

No horizontal linearity adjustments are used in the latest TV chassis, although in earlier TVs, horizontal linearity and width controls or coils are adjusted. In Fig. 6-14A, the picture width is pulled in on the right side. The width is adjusted too wide making a elongated picture in Fig. 6-14B. A normal picture is shown in Fig. 6-14C.

Horizontal drive

Although the horizontal drive adjustment in some TVs is neither a width nor linearity control, it can affect both; therefore, it should not be neglected during width or linearity adjustments. If the suggested procedures for correction of nonlinearity have been exhausted without achieving acceptable linearity, a check of the horizontal drive adjustment should be made. This is invariably a shaft-type adjustment, having a maximum rotation of less than one full turn.

After noting the original setting, the control is gradually rotated in one direction while you observe the effects on picture linearity, as described previously. That is, with the picture not quite reaching the edges of the screen. If the linearity worsens, the control should be turned in the opposite direction until effective correction is achieved. It is quite possible that, in an attempt to obtain some additional width, the

A

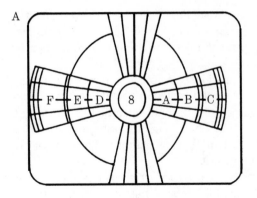

B

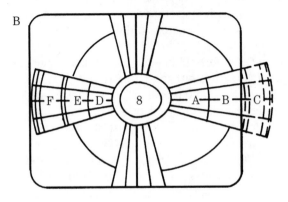

C

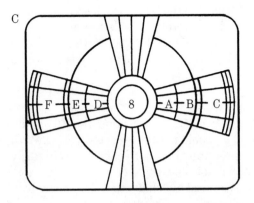

6-14 The screen is pulled in at the right side
(A), over scanned or too much width (B), and
the correct test pattern (C).

horizontal drive control was advanced to the point of overloading, distortion, and nonlinearity. The correct setting is that which produces the best linearity at a normal picture width.

Pincushion correction

During picture-linearity adjustments, it might be necessary to manipulate the pincushion correction magnets, a part of the picture tube neck assembly in some sets. Pincushion magnets are usually sliding metal tabs or bars and are intended to correct for some inherent design characteristics of the picture-deflection system. Looking at the back of the tube, a collar-like assembly, called a *yoke*, is positioned at the far front of the tube neck (see Fig. 6-15). The function of the yoke is to deflect (it is called a *deflection yoke*) the picture to a perfect rectangular format. Actually, however, and especially in large-size picture tubes, some distortion of the format, called *pincushioning*, occurs. The picture seems to be stretched as by pulling on the four corners, leaving the sides somewhat caved in (see Fig. 6-16).

Pincushion tabs or bars are fastened to the yoke structure and can be loosened for sliding, after which they are tightened again. As in other cases of nonlinearity described previously, these adjustments are best made when edges of the picture are visible.

A note of caution: The tube and deflection yoke assembly are rigidly secured. The pincushion correction was made during the installation of the original tube. It is usually not necessary nor advisable to make pincushion adjustments, unless there is clear evidence of distortion or when the picture tube is replaced (by a professional TV servicer, of course).

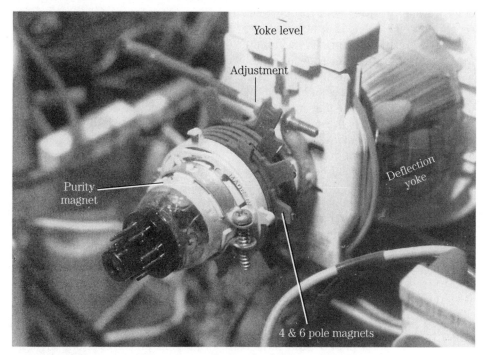

6-15 The different components on the neck of the picture tube.

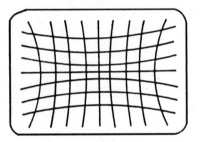

6-16
An improper adjustment or
defective pincushion circuits.

In some TVs, the pincushion correction is entirely automatic by virtue of a sensing, feedback, and compensating circuit. Because this is very reliable, there is little likelihood of malfunction during the life of the set. In case of failure, and after being sure that the distortion is not simple vertical or horizontal nonlinearity, the appropriate module board might need to be replaced. However, even in some of the so-called automatic self-correcting circuits in some recent sets, there is a manual presetting adjustment, somewhat akin to the presetting of the vertical hold, from which the automatic correction circuitry takes over.

Side pincushioning circuits

In early transistorized TVs and today's IC color chassis, both horizontal and vertical side pincushioning were corrected with a pincushion transformer. To compensate for pincushion distortion, the horizontal deflection current is modulated by a waveform during the vertical scan period. The inductance generated in the secondary winding varies with the current applied in the primary (Fig. 6-17). The secondary coil winding is in series with the horizontal deflection yoke. The vertical output is applied to the primary winding. When the parabola waveform and dc current is applied to the primary winding, a linear current is applied to the secondary winding, eliminating side cushioning in the raster.

With today's 27-inch color chassis, the pincushion-correction circuit might modulate the horizontal yoke current at a vertical rate to eliminate distortion. The pincushion transformer is in series with the horizontal yoke. The width changes with regulated B+ and modulated current through T502. As more B+ voltage appears across the yoke, the width increases.

The correction circuit develops a parabola waveform applied to the pincushion transformer (Fig. 6-18). Adjustment of the amplitude control determines how much vertical waveform is applied to the error amp. The beam current input to Q401 allows the pin-correction circuit to compensate for changes in width. The pincushion-correction circuit interacts with both the power source and vertical circuit.

Large-screen pincushion problems

The pincushion circuits with large picture tubes prevent the picture at the ends from turning up or down, or bowing inward. The pincushion circuit keeps the scanning lines linear and flat—even at the extreme corners of the picture tube. Sometimes it is difficult to see a picture that might have poor linearity at the corners and extreme edges of the picture tube. If the vertical sides of a building are bowed, problems exists in the pincushion circuits (Fig. 6-19).

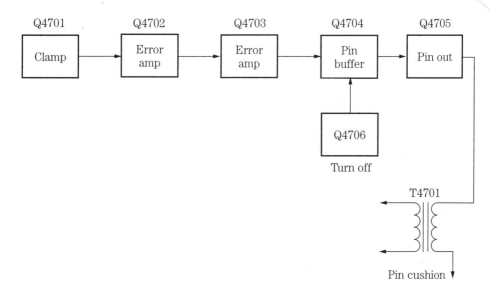

6-17 A block diagram of a pincushion transformer connected to the vertical and horizontal yoke winding.

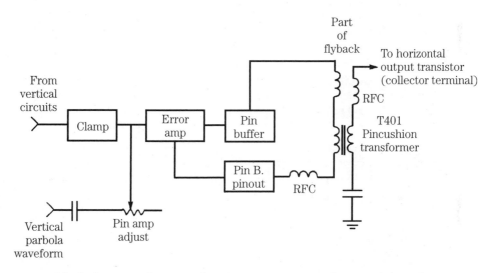

6-18 A block diagram of a pincushion circuit in a recently manufactured TV.

6-19
Bowed outside edges with a defective pincushion output transistor in a 27-inch picture tube.

The pincushion circuits in a 27-inch or larger screen might have a pin amp, pin buffer, and a pin output transistor to provide extended side adjustments, width, top-and-bottom, and phase adjustments. Often the pin output transistor becomes leaky and defeats the pincushion circuits (Fig. 6-20). In this case, extreme bowing is found at each edge of the picture and sometimes a hump is in the middle of the raster. After replacing the pin output transistor, touch up all controls in the pincushion circuits with a crosshatch pattern from the color dot-bar generator.

Centering

Another adjustment related to the various dimensional corrections is *centering*. The centering adjustments position the picture so that it extends equally in all four directions. If the vertical size is adjusted, both the top and bottom edges should be reached simultaneously if the picture is centered; likewise, when the width is adjusted, both sides should expand at the same time. In very old TVs, but also in some recent color TV sets, two separate shaft adjustments were used, one marked V-cent, the other H-cent.

Centering adjustments are very simple; the controls have a total travel of less than 360 degrees (less than one full turn) and can safely be turned back and forth until proper picture centering is achieved. In some TVs, the centering adjustments (if any) are mechanical/magnetic; the centering depends primarily on the original positioning and assembly of the deflection and focus components on the neck of the picture tube and only secondarily on after-the-fact adjustments.

The most popular type of centering consists of a pair of tabs, flat metal, magnetic strips, or bars, which can be rotated toward and away from each other until centering is achieved. One tab primarily affects the horizontal positioning, the other affects

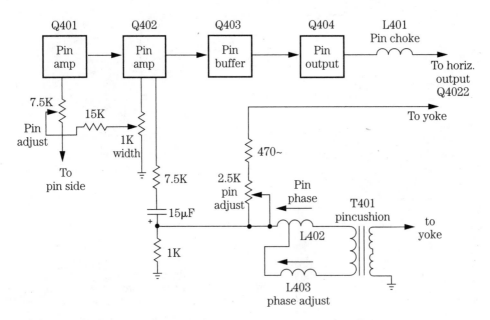

6-20 A block diagram of a pincushion circuit in a 27-inch color TV.

the vertical. Again, these are rather interdependent adjustments, and should, therefore, be made simultaneously. A variation of the latter type is the single-tab centering device. It can be moved both horizontally and vertically (in fact, it can be made to move in various directions simply by pushing down on it and sliding it sideways at the same time). Still another centering adjustment is a thumbscrew on the CRT neck components. This type requires only a clockwise or counterclockwise rotation to achieve centering.

Large-screen horizontal centering

The centering control is across the horizontal yoke windings. The only adjustable part in the horizontal-centering circuits is a horizontal-centering control (Fig. 6-21). All fixed capacitors should be replaced with 1-kV components. Adjust the horizontal-centering control with a test pattern or crosshatch-pattern generator.

Focusing

The focus adjustment varies the sharpness of the picture detail. As in the case of the other adjustments, there are different ways to achieve the same end result. Different manufacturers use different methods, but the procedures given here apply:

- *No adjustment*   A number of TVs have what is called *fixed focus*; the picture tube is designed and constructed so that the picture is always in focus, although there might be a provision for correction (by a servicer) in the rare case that this type of control is out of focus.
- *Focus thumbscrew*   Some focus controls are in the form of a short, flexible steel cable with a thumbscrew at the end; the cable extends through the back cover of the TV set. Rotating the knurled end of the cable adjusts the sharpness of the picture. Sets with this type hopefully have long since been put to rest, but once in a while you'll see one.
- *Sleeve adjustment*   This focus-correction device is in the form of a ring or sleeve on the neck of the picture tube. Sliding the ring or sleeve slowly along the neck of the tube adjusts the focus of the image. TVs in which the width control is also a sleeve-type adjustment, the width adjustment is very much forward on the picture tube neck, and the focus sleeve is very close to the rear end of the picture tube.
- *Shaft focus adjustment*   This control is similar to a volume control on the panel of an ordinary radio receiver. This type of control is used in some (usually older) black-and-white TV sets and on practically all color TVs, where focusing is very much of a requirement.
- *Picture tube focus adjustment*   You might find the focus control and screen controls attached to a plastic assembly that is plugged into the rear of the picture tube socket. In many of the recent RCA TVs, this plastic assembly must be unplugged before the picture-tube socket can be removed (Fig. 6-22). Some of the focus controls are screwdriver adjustments, and others require a hexagon alignment tool to rotate the plastic control. If the focus control has little or no effect on the raster, suspect that the picture tube is defective.

In all focusing adjustments, it is essential that a picture be watched at close range to obtain the sharpest focus possible. Although any fine detail of the picture

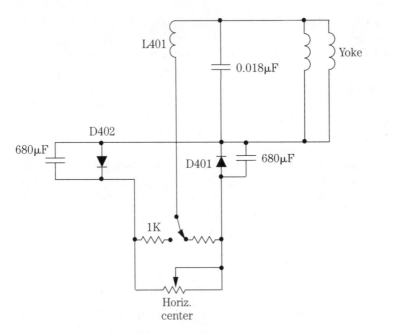

6-21 The horizontal centering circuit for a large-screen TV.

6-22 The focus and screen grid adjustments in a separate resistor network.

can be used to observe the effect of the focusing adjustment, it is best to use the horizontal lines, preferably on a bright portion of a scene, and focus them for the thinnest possible lines with the clearest separation.

Sub-brightness control

Turn the set on and let it warm up for at least 10 minutes before making any adjustments. Tune in a color program. Adjust the contrast control to minimum and the brightness control to maximum. Set the color and tint controls to the center of the rotation. Turn the sub-brightness control to the center. Some TVs are adjusted for a low-voltage measurement of the sub-brightness control or for a certain luminance measurement (100 to 150 Lux). Usually, the sub-brightness control is in the same circuit as the regular brightness control (Fig. 6-23). Be sure that the brightness can be turned all the way down to the minimum position of both the sub-brightness and regular brightness controls.

Arcing in the CRT socket

Continuous arcing in the CRT socket or focus control might cause horizontal firing lines in the picture. Excessive dust collected in the spark gaps of the picture tube socket might cause the TV to shut down. In early color TVs, the focus pin and socket formed green arcing residue that would sometimes arc over. Sometimes the socket or just the focus pin was replaced.

In the latest portable TVs, the spark gaps in the CRT socket might arc over and cause intermittent arcing or chassis shutdown (Fig. 6-24). The raster might come up, flash, dim down, then shut down. The spark gaps are located on the lock color amp cathode, grid, screen, and focus pin. Often, dust will filter through the air slots in the plastic cover and collect on the picture tube socket. Just blow the dust out or

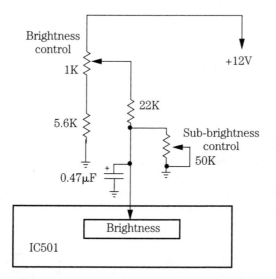

6-23 The sub-brightness control is in the same brightness circuit as the brightness control.

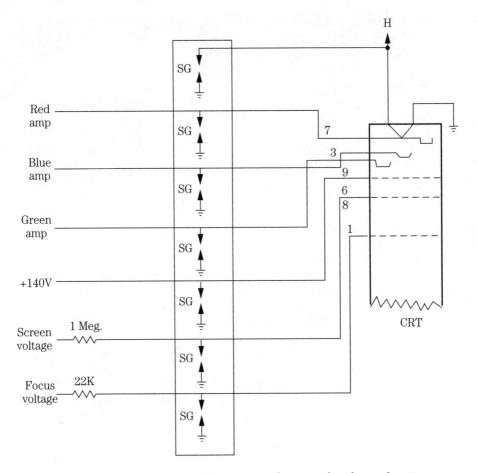

6-24 Arcover in the spark gaps of the CRT socket can shutdown the TV.

remove the socket from the CRT and tap it with a screwdriver handle several times to dislodge dust particles.

Arcing at the anode button

Excessive high voltage might cause the high voltage to arc over at the anode connection of the CRT. A collection of dust, moisture, dirt, and smoke can cause the high voltage to arc over from the anode button to the aquadag coating on the outside of the picture tube. Discharge the anode button with two large screwdrivers between the anode and outside of the aquadag coating of CRT. Do not discharge the picture tube between the anode button and the chassis ground or you might damage semiconductors in other circuits.

Remove the rubber cover and HV lead. Check the rubber area for arcing marks, cracking of the rubber cover, and excess dirt (Fig. 6-25). Clean it with cleaning fluid and let dry. Wipe off the button and the glass area. If the rubber cover continues to arc over or the HV cable is cracked, replace it with a new HV cable. If the HV cable comes directly out of the flyback, the horizontal output

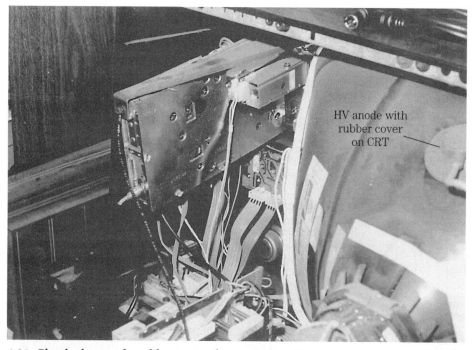

HV anode with
rubber cover
on CRT

6-25 Check the anode rubber cover for excessive dirt and dust causing arcing at the CRT.

transformer might need to be replaced because you cannot splice the HV cable successfully. These rubber connections can be replaced with new rubber plug, if only the rubber plug is damaged. Be sure that the plug is tight in the socket before firing up the chassis. Keep your hands away from the plug and the HV cable when TV set is turned on.

Ion trap

Some very old TVs have an additional picture adjustment on the back of the TV tube neck—the ion trap magnet. Because this can seriously affect picture brightness, size, and picture tube life, and because its position near the picture tube socket makes it vulnerable to accidental dislocation, it is important to consider its purpose and indicate the method of adjusting it if and when required.

The ion trap magnet is a form of bar magnet held on the picture tube neck by a flexible coil spring. Its purpose (in nontechnical language) is to ensure that the electron beam inside the neck of the tube is on course—moving properly along the axis of the tube. It is required because other electronic particles (ions) are incidentally generated inside the tube. Ions can damage a picture tube. The ion trap is thus part of a scheme of "separating the sheep from the goats"—getting rid of the ions, then being sure that the electrons move where they should be moving—through a number of small apertures toward the picture tube screen.

When adjusting the ion trap magnet, a combined sliding/rotating motion is used. The object is to get a dual result—full picture-screen coverage at maximum brightness. It is actually possible, through misadjustment, to illuminate only the central (circular) portion of the TV screen at a low brightness level.

Because ion trap adjustments can be detrimental to the picture tube, this procedure should be done carefully, repeated a few times, if necessary, until the best possible results are obtained. Although a blank screen (no picture) is easiest to observe during this adjustment, you can do it with any type of program on the screen.

In connection with the adjustment of the ion trap magnet, a note of caution is required. An improperly adjusted ion trap magnet, in addition to producing a poor picture, can also damage the physical interior structure of the picture tube, which will ultimately result in poor focus and dark spots (burn spots) at the center of the tube screen. To avoid this, the magnet should be kept near the base end of the picture tube while you very quickly make adjustments at the lowest possible brightness.

For easiest adjustment and best results, you should move the ion trap magnet as far back (toward the tube base) along the neck as possible, consistent with obtaining the desired results.

The brightness and contrast controls must be kept low so that slight changes are easiest to detect. After obtaining the best adjustment, it might be desirable to note or mark the position of the magnet both fore-and-aft, as well as around the neck, with a marking pen or a small piece of masking tape. Thus, if this adjustment is accidentally disturbed, incident to other work in the back of the TV set, it is necessary to move or turn the ion trap magnet to the marked location to restore the correct operation.

Ion traps haven't been used with new picture tubes in some years, but from time to time you might see an older set still operating with the original picture tube. New designs of picture tubes permit the deletions of the ion trap. If you own an old TV with a picture tube using an ion trap and purchase a replacement tube, you'll find that you can throw away the ion trap. The new tubes don't need it.

Magnet precautions

The following warning pertains to the effect of magnets, magnetic materials, and magnetic distortion on picture tubes. Although the subject of stray, unwanted magnetic effects on TV pictures came to the attention of the public only since the advent of color TV, the problem has always existed, but to a much lesser degree in other TVs. This subject, in connection with the matter of color purity in color TV sets, is covered later. This section is concerned with black-and-white TVs and how they can be affected by stray magnetism because of the careless handling of magnets and magnetic materials in certain areas of the TV.

As explained earlier, a number of adjustments, such as width, linearity, etc., depend either on the use of a magnetic material, or on the modification of an existing magnetic structure on or near the picture tube. Because iron and steel are magnetic materials (even if they are not magnetized, that is, they do not seem to attract nails, paper clips, etc.), they might, when brought near an existing circuit, change that circuit, at least temporarily. Because screwdrivers, pliers, wrenches, etc., are almost always constructed of iron or steel, their careless use on TVs could adversely affect the TV. For example, using a screwdriver blade to nudge or push an ion trap magnet or centering tabs is, therefore, not advised.

Similarly, using a pair of pliers to turn a tight thumbscrew, as used in some focus adjustments, can do more harm than good, particularly because the effect is not catastrophic, but relatively minor. Of course, it is quite proper to use such tools on the chassis and other necessary areas in the cabinet, provided that these areas are not too close to the tube neck, yoke assembly, focus magnet, or the adjustments on the neck of the picture tube.

Removing the magnetic field from the TV screen

A degaussing coil is mounted around the picture tube so that external degaussing after moving the TV set is normally unnecessary, provided that the receiver is properly degaussed on installation. This degaussing coil is only on for about 1 or 2 seconds after power to the TV is switched on. If the set is moved or faced in another direction, the power switch must be switched off at least 10 minutes so that the automatic degaussing circuit operates (Fig. 6-26).

If the chassis or parts of the cabinet become magnetized and cause poor purity, use an external degaussing coil. If the magnetic impurities stays on the screen at all times, suspect that a thermistor is defective or that a coil has shorted against a metal frame or cover. Sometimes these thermistor parts can be damaged by lightning or power-line surges and arc over the thermistor, leaving the degaussing coil on all of the time.

Other horizontal defects

Under the category of other horizontal defects comes such abnormalities as:

- *Tearing*   Picture breakup into horizontal strips shifted to left or right.
- *Shifting*   What seems like a vertical split down the middle of the picture with

6-26 Check the degaussing coil circuits if the coil is energized all of the time.

the right-hand half of the left and vice versa, often accomplished by a very thin, faint picture (see Fig. 6-27).

- *Snaking* The sides of the picture become wavy instead of vertical. This can also be accompanied at times by a rather thin, faint picture, although a normal picture might also be affected by this. See Fig. 6-28 for a typical example.
- *Streaking* The picture seems normal in every way, except for dashed dark lines that seem to streak across the screen from left to right, somewhat like Fig. 6-29.

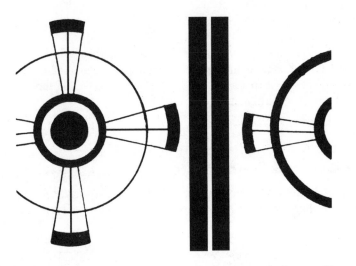

6-27 The horizontal circuits shifting or drifting off frequency.

6-28 A type of picture distortion called *snaking*.

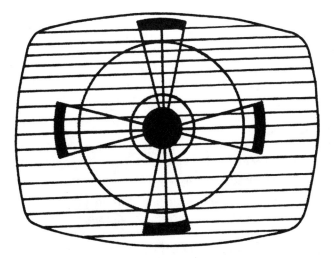

6-29 Picture streaking can be caused by human-made noise, motors, neon lights, and HV transformers.

Overloading

Paradoxical as this might seem, too strong a signal can be just as detrimental to normal TV performance as too weak a signal. Specifically, overloading because of too strong a signal can cause snaking and picture tearing, and is usually caused by one or both of the following:

- *Contrast control*   Picture tearing can often result from excessive picture signal. Incredible as this might seem, the technical structure of the average TV set is such that increasing the picture or contrast control to produce a very strong picture beyond that required for normal picture-viewing actually decreases the strength of the sync signal and often results in tearing. If this is the case, turning the same control toward a more moderate contrast setting should correct the tearing.
- *AGC pots*   Most TVs manufactured in the past few years have a rear-chassis adjustment called *AGC (automatic gain control)*. Its function, when properly set, is to place the receiver in such an operating condition that all stations in the receiving area will produce an acceptable picture with only an occasional adjustment of the front-panel contrast control. However, misadjusting the AGC pot can cause TV woes, not the least of which is loss of picture stability (hum, buzz, and even a complete loss of picture). Because most TVs have a tube whose function is to regulate AGC performance, a defect in the tube could also be responsible. Because of these possibilities, the following procedure applies to the AGC stage and the AGC pot setting.
- *AGC adjustment*   Using a moderately strong TV station, set the contrast (or picture) control about halfway, or a bit beyond (clockwise). Do not back off the contrast—even if the picture seems too strong. While watching the picture in a mirror, advance (clockwise) the AGC adjustment on the back of

the chassis until a buzz (or hum) is clearly heard. In some sets, advancing the control too far beyond this point might cause a complete loss of picture. Exercise moderation. Back off the AGC adjustment until all traces of the buzz disappear, or just slightly more than that. The AGC is now correctly set. Any malfunction of the horizontal sync will no longer be caused by the AGC setting, and other corrections must be attempted.

Horizontal hold, lock, and range

Although all three of these adjustments might not be used in any one set, all have a horizontal hold adjustment, and some have one or the other of the remaining two. Any one of these might be responsible for picture tearing or instability:

- *Horizontal hold*   Assuming that the tube is good, a picture might fail to hold if the horizontal hold control is at either extreme of its range. It is permissible for the picture to begin to show tendencies of tearing or instability when the control is at either end of its travel. As mentioned earlier, some horizontal hold controls have less than a full turn of adjustment range. In that case, a fraction of a turn is all that will be required for correction. In slug-tuned horizontal hold controls, two or three turns in one direction or the other should correct any instability because of this setting.
- *Horizontal range and lock*   These back-of-the-set adjustments are intended to set the range over which the automatic horizontal synchronization is effective. The proper setting is that which will produce no loss in horizontal stability when adjusting the fine-tuning control (if any) or when switching from station to station. Starting with a condition of instability or tearing, the horizontal range or horizontal lock control is adjusted very gradually until the tearing disappears (picture is normal). Then a test for stability is made by switching channels or by turning the front-panel hold control to near extremes (almost fully clockwise, then almost fully counterclockwise). In either case, there should not be a loss of sync.

Color horizontal adjustments

In early transistorized color chassis, the horizontal oscillator could be adjusted by rotating the ceramic metal core inside the horizontal oscillator coil. Within later horizontal circuits, they were located inside a deflection IC (Fig. 6-30). Instead of a horizontal oscillator coil, R607 varies the voltage applied to the internal horizontal circuits, locking in the picture. You might find one large IC processor that contains all the deflection circuits plus the sync separator, AGC and ABC detector, IF processing, and sand-castle generator.

No vertical or horizontal controls

You will not see any outside horizontal or vertical controls in many of the new color TVs. Both the horizontal and vertical oscillator stages are synchronized internally with the incoming signal. The horizontal oscillator circuits can be found in one IC developing the deflection signals, sand-castle output pulse, coincidence voltage output, and video IF processor (Fig. 6-31). The horizontal oscillator begins oscillating when voltage is applied to pin 10. The horizontal oscillator is locked in with the

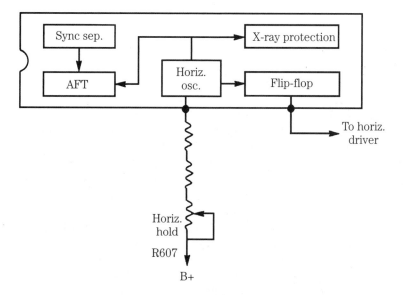

6-30 A block diagram of the horizontal circuits inside an IC.

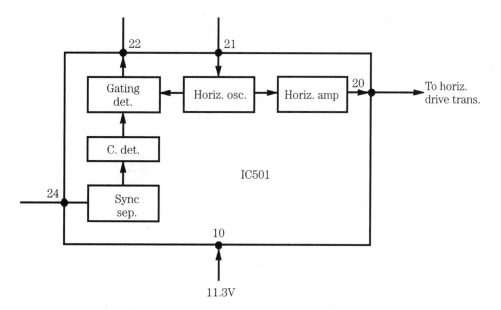

6-31 A block diagram of the horizontal circuits in a deflection IC.

incoming signal at pin 24. The sync separator circuit provides signal to the horizontal sync signals to the gating and phase one detector. The horizontal oscillator frequency is controlled by a dc voltage at pins 22 and 21. The horizontal oscillator drives the horizontal amp to the horizontal driver stage.

The horizontal drive circuits without horizontal hold controls might be an IC (which includes deflection circuits, AGC, IF amp, video detector, AFT, sound IF, audio processing, luma/chroma processing, sync separator, X-ray protector, horizontal AFC, VCO and horizontal countdown). The horizontal countdown signal generates the drive signal for the horizontal drive transistor.

The frequency of the horizontal oscillator is controlled by the VCO crystal (Fig. 6-32). The output of the VCO circuit is applied to the horizontal countdown circuit. The countdown circuit divides the signal to the proper frequency for the horizontal deflection circuit. The horizontal signal is applied from a pin to a buffer and the horizontal drive transistor. The crystal-controlled horizontal circuits provide stable operation without a horizontal hold control.

Streaking

This display is unmistakable and it is caused by an electrical noise (commonly called *static* in radio) of the manmade variety, and there is no simple way to eliminate it. It might appear either as long dashed lines traveling slowly across the screen, or as a random band or stream of snow, also moving across the screen in unison. If it is chronic, it might require relocation of the TV antenna by trial and error—a very costly remedy.

The noisy condition usually occurs from channel 6 on down. When channels 12 and 13 have excessive noise, you can assume the noisy system is close by. You can

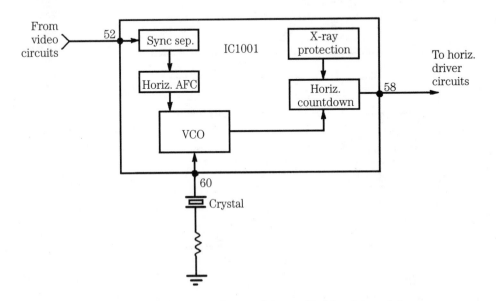

6-32 A block diagram with internal crystal-controlled horizontal frequency of an RCA color TV.

check for excessive noise coming into the power line of the house with a portable radio. If the noise is not noticeable at the power-line entrance, suspect that the noise is being received by the outside antenna.

If the noise can be traced to its source (for example, an arcing power company pole transformer, a heating system motor that sparks excessively, a similar workbench motor, or some other defective electrical appliance), the proper procedure is to correct the condition at its point of origin. If the source is a utility company device, their cooperation will be required. It can usually be obtained, provided that a company representative is shown the evidence. There should be no difficulty in recognizing the malfunction and, if it is his company's responsibility, he will arrange for correction. If an in-house appliance is causing the noise, correction will have to be made by a competent servicer familiar with such appliances (oil burner, for example). More on this subject in the next chapter.

Intermittent streaking and picture breakup

The tuner or front end can cause erratic, intermittent streaking, rolling, and tearing of the TV picture. This very critical and most sensitive of all sections of the TV (black and white or color) is, without question, the mostmanipulated and most worn portion of the set. Regardless of make or model, the tuner contains a large number (up to 100 in some sets) of delicate switches or contacts, undergoing make-and-break operation whenever a station is changed. Because, as explained earlier, the precise sync-timing signals are an integral part of the TV picture, the signal interruption caused by switching from channel to channel also removes the sync pulses. For a moment, the TV is on its own.

In a normally functioning TV tuner, that is, one that is not worn out or otherwise defective, switching channels momentarily interrupts the timing signals, but they are restored almost instantly. During the switching interval, a quick tear or roll might occur, or there might be a momentary diagonal zig-zag. One frame (picture) might slide up or down. In a worn tuner, including virtually every tuner after a few years use, the switch contacts start to become intermittent and erratic, causing some misbehavior of the kind just described.

With a color receiver, malfunctions (such as color confetti, absence of one or more colors, and especially color instability and shifting) are often caused by poor contacts in the tuner. No special point is made of this fact under the color picture troubleshooting because the problem is not peculiar to color. In fact, it is much preferable (as is indicated in the introduction to color TV troubleshooting) to cut off the color when troubleshooting such a problem. When operation is restored to normal on monochrome, color performance will automatically have been corrected. In other words, this is a basic receiver problem, not a color problem.

There are two ways to solve a noisy or intermittent tuner problem: either thoroughly clean and lubricate it or have it overhauled by a specialist. A noisy tuner is so common in TVs that a very practical and relatively inexpensive procedure has been established and is being followed by the repair industry. The old tuner is sent to a specialist overhaul shop where a factory-type overhaul is performed on each tuner. The end results might be as good as new. In view of the heavy use the tuner must endure, anything less than such a professional overhaul

6-33 The varactor tuner might drift off channel, and cause one or more channels to become noisy and snowy.

is a poor second choice, except where this service is not available or where a competent servicer is certain, after examination, that the overhaul is not required (Fig. 6-33).

If you decide that cleaning and lubricating is the best course, you should begin, if you feel confident enough to tackle the job yourself, with a detailed examination of the before conditions, including the method used to attach the tuner to the front panel or the chassis, and the exact positions, color, marking, etc., of each lead and cable from the tuner to the chassis. Most leads in newer sets are of the quick-disconnect type, but if one (especially wire braid or shielding) is soldered, carefully unsolder it, using no more heat than necessary. When you reassemble, you must be just as careful when resoldering to the same spot. Be sure that it isn't a cold-solder joint (a seemingly rigid mechanical joint with a grainy appearance that will pull off if tugged hard or pried). It is very important to return all wires to the same locations as before.

After removing the tuner from the chassis, look at the contacts. Misalignment and looseness should be apparent now. In the case of snap-in channel coils, remove one coil to observe the interior. Using a fine, soft brush (a small camel's hair artist's brush is best), clean the wiping contacts and the shaft bearing with a contact cleaner (sold in electronics stores). Do not use household grease solvents— especially not carbon tetrachloride. Allow all surfaces to dry. The contacts are best cleaned while repeatedly switching channels back and forth.

CRT gray-scale adjustment

Set the color control to minimum. Readjust the brightness and contrast controls to obtain a low-light area. Now adjust red, green, and blue bias controls to obtain a gray raster of low brightness (Fig. 6-34). Adjust the brightness and contrast control to maximum. Adjust the blue and red drive controls to obtain power white-balanced picture in high-light areas. Go over the adjustments of brightness, contrast, and drive controls for a correct gray scale.

HV adjustment

Check the HV before making any adjustments or repairs of the TV chassis. Some manufacturers might call this the B+ voltage adjustment. Actually, the high-voltage adjustment varies the dc voltage applied to the horizontal output transistor. If this voltage is too high, the high voltage might shut down the TV chassis. You might find there are no adjustments with some TVs to vary the HV at the picture tube. Always check the B+ setting after repairs to prevent possible arc over damage.

Connect the high voltage meter probe to the anode button of the picture tube. Turn the chassis on, set the brightness and contrast controls to minimum position. Does the high-voltage reading compare with the voltage on the schematic? Likewise, check the correct voltage source that feeds the horizontal output transistor. Be sure that the ac power line voltage is around 120 volts. Adjust the HV or B+ control until the correct B+ and HV are measured at the CRT (Fig. 6-35). Rotate the brightness control be sure that the high voltage does not exceed 5 kV over the normal listing. If the HV is too high and the chassis shuts down, check the horizontal safety capacitors and B+ voltage.

Tuner spray lubricant

There is a quicker, although not necessarily better, way to correct noise and poor contacts in the tuner without resorting to the more tedious removal, cleaning,

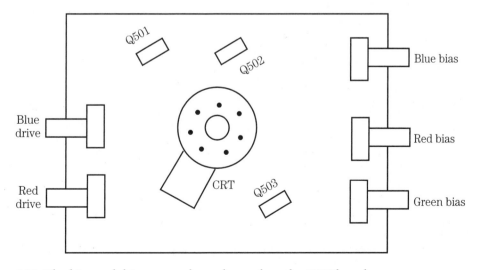

6-34 The bias and drive controls are located on the CRT board.

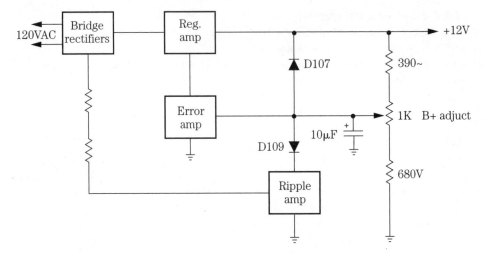

6-35 The high-voltage or B+ adjustment should be adjusted to the correct B+ source.

and lubricating procedure. It is not a cure, but a correction that might last for quite a long time. Spray lubricants are used by many TV servicers whenever it is decided not to remove or replace the tuner.

Electronic component supply houses sell a pressurized spray cleaner/lubricant, marketed under various names, usually described as a cure-all or similar solution. The procedure still requires removing the TV from the cabinet and removing the protective shield from the tuner. A small amount (so as not to cause dripping) is sprayed onto the contacts of the tuner switch sections while the station selector is manipulated, as when changing stations, to enable the fluid to reach all contact surfaces. The solution dries quickly and is claimed to leave no objectionable residue, only a film of conducting lubricant. Incidentally, removing the selector knob and spraying into the tuner, as suggested by one spray manufacturer, is hardly a satisfactory solution.

Tuner wash

Many different types of cleaning chemicals are available to clean the tuner contacts. In some wafer-type tuners, excess grease attracts particles of dust that become hard and result in a very dirty tuner. These tuners might not become clean with an ordinary tuner cleaner. A large spray can equipped with additional plastic tubing and capable of delivering a heavy spray of cleaning fluid should be used to flush out the lumps of grease and dirt.

Place a drip cloth under the tuner and protect the chassis area from the washing liquid. Place the small plastic end right on each wafer switch contact assembly. Spray both sides of the wafer switch to dislodge all greasy residue. Rotate the tuner switch assembly so that each section can be washed out. After applying several coats, complete the tuner clean up by spraying the wafer switch contacts with regular spray lubricant.

Don't just turn any screw

Often, when the TV set acts up, a person has the tendency to turn a few controls to get the program back on the screen. Readjustment of all external controls are placed conveniently for this purpose. Do not go inside and start turning a bunch of preset controls. Many screws and preset adjustments inside the TV should not be touched, unless you have the correct equipment for alignment and making certain tests (Fig. 6-36).

Leveling the picture

If the picture is not level, loosen yoke bolt with a ¼-inch nut driver. This metal-type screw is on a metal clamp at the rear of the picture tube (Fig. 6-37). Just loosen the screw and rotate the yoke until it is level. Hold a mirror in front of the TV screen so that you know when the picture is level. Now, take a look at the front of the TV. Lock the metal screw with a nut driver.

6-36 Do not adjust any controls on the TV chassis without correct test equipment and knowledge.

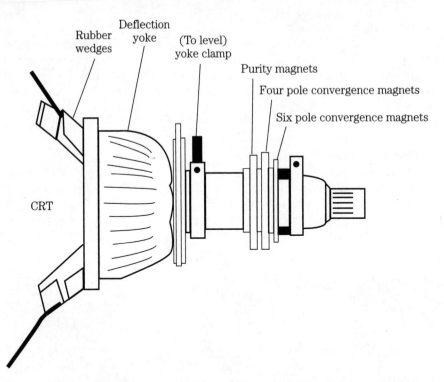

6-37 Level the picture by rotating the yoke assembly on the picture tube.

7

CHAPTER

Internal troubleshooting

Chapter 6 covered what might be called "external troubleshooting"—the problems and solutions associated with the various controls and adjustments accessible from the outside of the TV. This chapter and Chapter 8 concentrate on defects and cures that originate beyond these adjustments in the interior of the TV. Because there is no absolutely clear dividing line between these origins and locations of defects, there will necessarily be an overlap between them; however, it is desirable to separate these because the so-called "external" or adjustment type of defects are the most common.

Blooming and defocusing

This phenomenon is common in older TVs. A picture with blooming and defocusing problems exhibits insufficient brightness and contrast. Advancing either control, but particularly the brightness control, seems to have the effect of defocusing the picture, softening it as if a cloud were spread over the picture, fuzzing over the sharp details. The brightness might increase somewhat, but not enough before detail, and to a certain extent, shape is degraded. The fault lies in either or both of the following two areas.

High-voltage supply and picture tube

There is a considerable amount of interaction between the high-voltage power supply and the picture tube. In an old picture tube, the electron emission capability might be substantially below normal. Increasing the brightness setting further taxes an already-depleted source of electrons; at the same time, it upsets the voltage distribution in the high-voltage (kilovolts) output. This is directly responsible for picture brightness. Similarly, the natural aging of some tubes in the horizontal deflection high-voltage system further acts to produce blooming and decrease electron beam acceleration.

High-voltage probe

If you plan to do any future TV servicing, the high-voltage probe is a fairly inexpensive test instrument to measure HV at the anode button of the picture tube (Fig. 7-1). Do not try to measure the focus or HV with an VOM or DMM. Choose a HV probe that measures up to 42 kVdc. Some of the early meters were calibrated at 25 kV. The high-voltage probe has a high multiplier resistance to safely measure high voltage in TVs.

Before inserting the HV probe under the anode rubber cover, be sure that the ground clip is clipped to the chassis or picture tube ground. Locate the ground on the outside aquadag coating of the CRT and ground it at this point. If not, you can receive a terrible shock; you might drop the HV meter, damaging the test instrument and other components nearby. Often, the HV probe can be inserted under the rubber cover and left there until the horizontal or high-voltage circuits are repaired.

Poor focus

Check for a defective tripler, horizontal output circuit, flyback (IHVT), and voltage-divider network for poor focus problems (Fig. 7-2). Measure the high voltage at the CRT and check it against the schematic. Measure the focus voltage applied to the focus pin terminal of CRT. This voltage should measure between 3.5 and 6.5 kV with the average picture tube and from 6 to 9 kV in larger size CRTs. Check the horizontal circuits if both the HV and focus voltages are very low.

If the high voltage is very low, suspect that a horizontal output component is defective. Check the voltage at the collector terminal of the horizontal output transistor or take a scope waveform. Measure the voltage from the dc source feeding the horizontal circuits. Likewise, adjust the HV or B+ control for low voltage. If the focus control has little effect on the focus of the picture, suspect that a focus-screen resistor network or CRT is defective. Test the picture tube with a CRT tester.

High-voltage rectifier

The high-voltage rectifier in the tube chassis was a diode tube with cathode and plate elements (such as a 1B3, 3A3, and 3AT2). In the early solid-state chassis, the high-voltage rectifier consisted of a tripler unit. Today, the high- voltage diodes are molded right inside the flyback, (IHVT), integrated HV transformer (Fig. 7-3). You

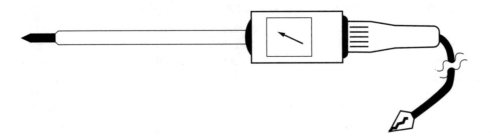

7-1 With a high-voltage probe, you can measure the HV at the anode button of the CRT.

7-2 Check the focus voltage at the CRT if the focus control has no effect on the picture.

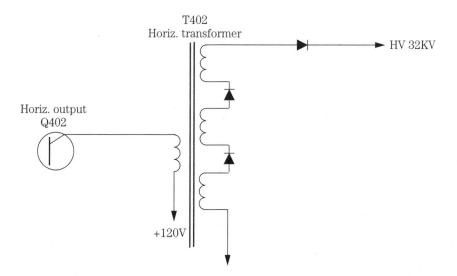

7-3 The integrated high-voltage transformer (IHVT) has diodes in the high-voltage winding.

cannot see these diodes, but you can hear a firing or cracking noise if they arc over. Replace the whole flyback with defective HV rectifiers. A defective horizontal output transformer can arc over internally, short between windings and transformer core area, and become leaky between HV diodes.

A continuity or resistance test of the primary winding shows continuity, except that it will not show shorted turns of wire. You cannot measure any continuity of the secondary winding because of the HV diodes in the circuit. The defective flyback might smoke, arc on the outside cover, and cause an audible internal arcing of the HV diodes. A quiet tic-tic noise does not necessarily indicate that the IHVT flyback is defective. All other horizontal output circuits must be checked and tested before you replace the horizontal output transformer. The HV rectifier in the black-and-white chassis consists of several layers of selenium, which forms a stick rectifier.

Damper diode

The damper diode often shorts or becomes leaky. In many cases, when the diode shorts out, the line fuse is blown. The damper diode does not act like the tube in the damper circuits for blooming conditions. Either the diode shorts or becomes leaky (Fig. 7-4).

The damper diode is located in the collector circuit of the horizontal output transistor. Within the latest TVs, the damper diode might be included with the horizontal output transistor. The diode connects from the collector to emitter the terminal inside the transistor. If the fuse in the solid-state chassis keeps blowing, check for

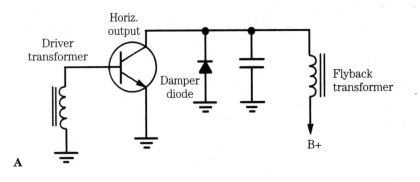

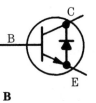

The damper diode outside of the horizontal output transistor (A) and inside transistor (B).

7-4 The damper diode might be outside the horizontal output transistor (A) or mounted internally (B).

a leaky damper diode or horizontal output transistor. Simply measure the resistance with the diode check of a DMM. Measure from the metal body of the output transistor to the chassis ground. A good diode and transistor will measure around 0.500 Ω in only one direction. If a measurement in both directions from the transistor body and chassis is 100 Ω or less, suspect that a damper diode or horizontal output transistor is leaky.

Horizontal output transistor

The horizontal output transistor is replaced more often than any other component in the TV chassis. This transistor can be located on a separate metal heatsink or chassis (Fig. 7-5). Some output transistors run warm. If it is red hot, suspect that a transistor is leaky or that the drive voltage at the base terminal is improper.

If you find a separate heatsink, no insulator is between the metal case of the transistor and the heatsink. But, when mounted on the metal chassis, the transistor is insulated away from the metal chassis with a mica insulator. Apply silicone grease to both sides of the mica insulator to help dissipate heat and prevent shorting the dc supply voltage. Use a clear silicone grease, instead of white, which is rather messy on your hands and clothes.

The defective horizontal output transistor might appear leaky, open, or shorted. A leaky output transistor can blow the main power fuse or trip the circuit breaker. This transistor usually becomes leaky between the collector and emitter terminals. Because the emitter terminal is grounded, take a quick resistance measurement between the metal body of the transistor (collector) and the chassis ground to determine if the transistor is leaky or shorted.

Remember, in some TV chassis, this transistor is mounted above ground or on a positive ground level (Fig. 7-6). Notice that the driver transistor, transformer, and horizontal output transistor are grounded to a hot ground. When taking voltage or scope measurements on these components, place the black (common) voltmeter probe to the hot ground terminal. If not, improper voltage, resistance, and waveforms will result.

Horizontal driver transistor

The horizontal driver transistor receives the horizontal drive pulse from the deflection IC and applies this signal to the base of the driver transistor. For instance, the driver transistor might receive a 4.5-volt pulse or waveform and amplify the signal to a 185-volt waveform. A horizontal driver transformer is located between the driver and horizontal output transistor for the correct impedance, insulation, and drive waveform (Fig. 7-7).

Check the collector voltage to indicate low voltage with an improper drive signal or a leaky transistor. The collector terminal receives a high dc voltage from the same source as the output transistor. A shorted or leaky driver transistor could overheat the primary winding of driver transformer and burn or overheat the isolation resistor in the voltage source. Improper or no input drive voltage might cause this driver transistor to overheat or become very warm. Remember, this transistor runs fairly warm with normal operation. Accurate voltage and waveforms taken at the terminals of the driver transistor can indicate that a transistor or stage is defective.

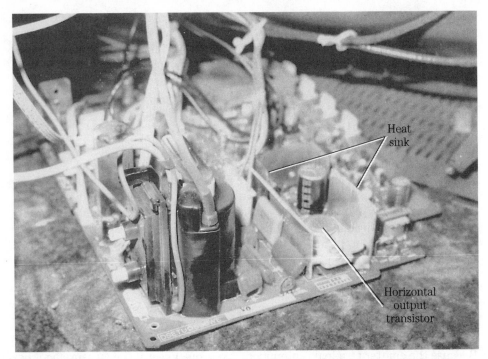

7-5 The horizontal output transistor can easily be located on a separate heatsink or on the metal TV chassis.

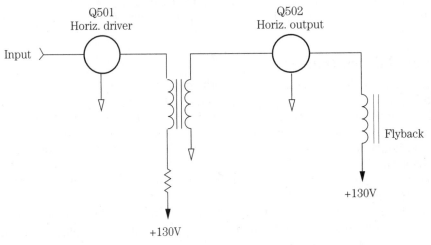

7-6 Some of the latest TVs have a hot ground in the horizontal output circuits.

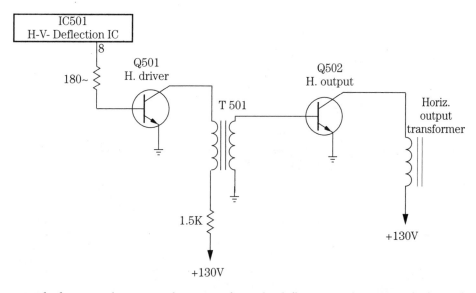

7-7 The horizontal output pulse is sent from the deflection IC (IC501) to the base of a horizontal drive transistor (Q501).

If the chassis shuts down, solder all terminals on the PC wiring of the drive transformer. Check the resistance of the primary winding and compare them to the schematic. This winding might have burned internally and shorted some turns if the driver transistor overheated. This transformer can prevent chassis startup and also chassis shutdown. Replace the driver transformer if it has burned or if the primary winding has changed in resistance.

Picture-tube problems

If too much brightness or contrast is raised, notice the fuzzy white, blotchy picture with a close up of a person's face. Often, this condition is caused by a weak picture-tube gun assembly (Fig. 7-8). If the color bleeds in sections, suspect that the CRT is weak and gassy. Also, a very weak picture tube results is not adequately bright.

A picture tube rejuvenator or booster will extend the remaining life of the tube by a number of months. Be careful to obtain the correct type of booster (for series or for parallel tube filament sets). Actually, the higher filament voltage results in more electrons flowing from the cathode and bouncing against the color beads on the front of the picture tube.

For three-gun color tubes, boosters have become available and can be used, provided that the instructions are carefully followed. Because the color tube consists of three separate electron-emitting structures, some boosters and rejuvenators have a simple provision for boosting one of the three, depending on which of the color guns is deficient, by externally adjusting the booster before used.

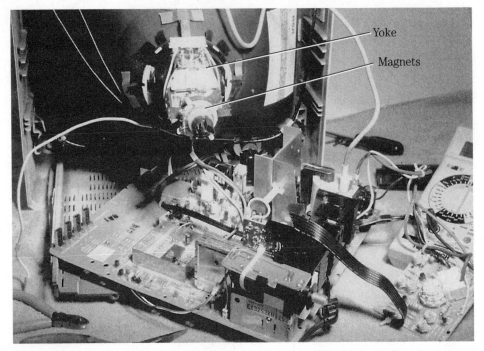

7-8 A weak gun assembly within the picture tube can result in a very dim picture.

Picture tube rejuvenation

Improper heater voltage or no voltage can produce a weak or entirely black screen. Of course, improper high or low voltage applied to the picture-tube elements can produce a weak or dead raster. Sometimes by peering down toward the picture tube socket, you may see all three heaters light in a color picture tube.

The color picture tube has three separate gun assemblies, including heater and cathode elements. The filament, like any tube, is enclosed inside a nickel-plated cathode that emits electrons. If the cathode is used up or is bombarded with layers of ions, it will emit fewer electrons. If the heater voltage is raised or the cathode is cleaned off of ions, the gun assembly will appear almost as good as new for a few months.

The electronic technician might have a color picture tube rejuvenator that restores the picture tube. The rejuvenator is enclosed in the same test instrument as the tube tester. This instrument will strip or clean off the ions from the cathode element, letting thousands of electrons to emit and strike the color beads on the glass front of the picture tube. The ions are a result of electrons bombarding the screen and falling back on the cathode, decreasing the flow of electrons (Fig. 7-9).

A high voltage is applied to the cathode element from the picture-tube rejuvenator when the heater voltage is raised, removing or cleaning up the cathode surface (Fig. 7-10). You can see the flashes of arcing and light around the cathode element when the rejuvenation occurs. Each color gun is cleaned and rejuvenated. Sometimes all three guns will be cleaned, restoring the picture tube, almost like new. Other times, one or two guns will not come up at all. Usually, there is a separate

7-9 Check the CRT with the tube tester and rejuvenate each color gun assembly with tube-rejuvenation test instrument.

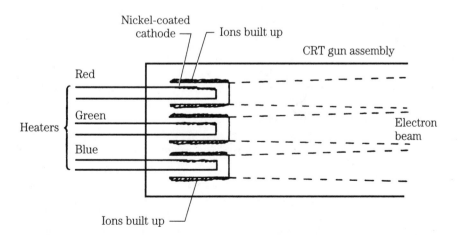

7-10 The tube-rejuvenator test instrument removes ions built on the cathode elements.

charge for rejuvenating a color picture tube. Sometimes the rejuvenated picture tube will last for the life of the TV. Other times, it will last only six months or more.

Picture tube damage

If you are not careful, working around the picture tube can be quite dangerous. The hazards are a possible imploded tube and high-voltage shocks. All picture tubes have an integral vacuum, so handle them with care. When taking the back off of the TV cabinet, do not accidentally drop the back cover on the neck of the picture tube and break it off. Not only is it very costly, but quite dangerous. Receiving a shock from the high voltage of the picture tube is not life-threatening; however, it might cause you to jump, fall, or accidentally pull test equipment off the service bench and damage other products nearby.

Always discharge the CRT before attempting to replace parts or work around the high-voltage circuits. The aquadag and the glass with the high voltage acts like a high-voltage capacitor and can hold a charge for several days. Discharge the anode button to the outside of aquadag before attempting to remove the chassis from the cabinet. Do not discharge the anode voltage to the common ground.

Replacing the picture tube

After making several tests on the picture tube and deciding that the tube is defective, discharge the HV, and remove the tube from the cabinet (Fig. 7-11). First, remove the chassis by loosening two side screws or bolts. Some chassis can be bolted from underneath the cabinet in console models. Disconnect all cables, the speaker, picture tube socket assembly, deflection yoke, degaussing coil, and ground wires. Now remove the chassis from the case or cabinet.

Place the cabinet or portable face down on a blanket or pad to protect the front of cabinet and picture tube. Remove metal screws or bolts at each corner of the tube. Remove degaussing coil assembly with holders at each corner. You might find plastic ties in each corner holding the degaussing coil (Fig. 7-12). Leave all components on the neck of the tube until the CRT is placed on the service bench.

Two people should be used to remove a large picture tube from the cabinet. It might take more help to remove a 27- or 31-inch picture tube. Set the tube face down on the bench cloth or pad so that you do not scratch the front screen. Always wear safety glasses or goggles when removing and replacing the picture tube.

Measure the distance from the end of picture tube to the tab magnets of the old tube to correctly reposition the replacement (Fig. 7-13). Remember that some deflection yokes are bonded to the neck of the tube and cannot be removed. In this case, order a new tube (with the correct part number) with a bonded yoke assembly. A new yoke assembly will come with the new replacement, if it is not interchangeable.

Replace all components on the neck of the new tube where they were mounted on the defective tube. Check the position of the purity and beam magnets on the neck of the new replacement (Fig. 7-14). Reverse the procedure when installing a new picture tube. Picture tube replacement, adjustments, and alignment must be correct. If you feel uncomfortable starting to replace the defective tube, then have the professional technician install it.

7-11 Remove the chassis and all cable connections before removing the picture tube from the cabinet.

7-12 Remove the chassis and cables from the portable TV before attempting to remove the defective picture tube.

7-13 Check the settings of the ion trap, blue lateral magnets, and convergence harness on older picture tubes.

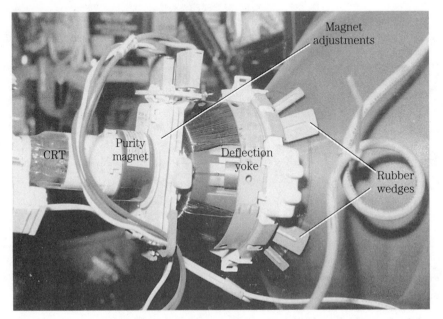

7-14 Measure the distance of the various magnet assemblies on the neck of the picture tube so that they can be replaced in the same position.

Dim and blooming when set is cold

In early TVs, before warmup, the tubes provided a dim and blooming picture. When the picture tube is weak, the picture lacks brightness. The picture can be improved by installing a picture tube brightener. The tube brightener raises the heater voltage, providing a greater release of electrons bombarding the screen. Although installing a brightener is only a temporarily repair of the CRT, it does result in a low-cost repair.

Attach the booster to the picture tube, following instructions given in Chapter 8. Allow the tube to operate with the booster for a few days. If the trouble disappears, the picture tube is at fault. Three remedies are possible:

- *Leave the booster on permanently* It will prolong the life of the dying tube.
- *Have the tube rejuvenated* Approximately the same results will be obtained, but the booster will not have to be used.
- *Have the picture tube replaced* Under normal conditions, between 80 percent and 90 percent of the tube's useful life has already been realized, so the cost of a new tube is quite justified if the TV is otherwise performing satisfactorily.

If the booster does not materially improve the performance, a check of one or two other tubes in the set should be made. This includes the high-voltage rectifier and the damper tube, in that order.

If none of these tips helps you to locate the cause, it is possible that a component (most likely a filter capacitor) in the high-voltage or B+ boost circuit is responsible. This, however, is not a task for a beginner.

Hum and buzz

These two names, *hum* and *buzz*, actually stand for two different symptoms. However, because it is sometimes be difficult for an inexperienced ear to clearly distinguish between the two, and because the two often appear together, I describe them together. However, the sources of these problems are not the same in all cases.

Hum

Hum can best be described as a smooth, steady low-pitched note that is the standard characteristic of household (60 hertz) ac. It can usually be heard on almost any tube-type radio with the volume control fully lowered if you place your ear close to the loudspeaker. It is almost always independent of the station being received, the weather, volume control setting, etc. In fact, it is the sound of 60-hertz ac. In most tube receivers (and sometimes in transistorized sets, when not operating on internal batteries), the hum is caused by a small residue of this ac sound that is not filtered (Fig. 7-15).

In severe cases, the appearance of hum signifies one of three possibilities: First and most likely is that one or more of the filter capacitors in the receiver (radio or TV) has deteriorated with age and heat so that it does not function at maximum effectiveness. Because there is no simple visual indication or identification of the defective component, the services of an experienced technician are required. A second, although less frequent, cause of hum is a defective tube, which

7-15 A dried-up electrolytic filter capacitor can cause hum in the speaker and dark bars in the raster.

has developed a partial short between the ac input and one of the functional elements. In technical parlance, this is usually called *heater-to-cathode leakage*. This can be analyzed, located and remedied by following a simple test procedure.

Turn the volume up or down. If this seems to make no difference in the hum level, turn the volume fully off. If no change in the hum results (there probably will be no change), the fault most likely lies on one of the filter capacitors. Incidentally, this type of defect seldom happens suddenly, it is a gradual deterioration, not a catastrophic failure.

If the hum level changes with changes in the volume control setting, suspect hum pickup in the AF stages, ahead of the volume control. If the hum remains with the volume turned way down, check the main filter capacitors in the low-voltage power supply. A dried-up electrolytic capacitor in the low-voltage source can cause poor width, black hum bars, relay chatter, and hum in the speaker. Excessive pulling and crawling of the picture can be caused by a defective filter capacitor.

Go directly to the main filter capacitor if one or two black bars is slowly rolling up the picture or raster. Sometimes black bars in the raster result from a shorted or leaky video transistor or IC. Hum bars in the picture are caused by a defective filter or decoupling capacitor in the low-voltage sources. Simply eliminate the video problem by turning the TV off channel or so that no picture is in the raster. If the dark lines are still located in the raster, with a loud hum in the speaker, check the main filter capacitors.

Other components can cause dark bars in the raster or picture with improper adjustment of the B+ or HV regulator control. Readjust the B+ control and notice if the dark bars disappear. Set the B+ control to the correct voltage applied to the horizontal output transistor. A leaky regulator transistor or corresponding zener diode can cause dark bars in the picture or raster. First check the main filter capacitors, then check the voltage regulator circuits.

For instance, if you do not have a matching replacement capacitor, install a higher-rated capacitor. If the old capacitor is 690 µF at 180 volts and you have a new replacement of 870 µF at 200 volts, clip it across the suspected one. If the dark bars disappear, replace the defective filter capacitor. Filter capacitors can be changed with units that have higher capacitance and working voltage. Do not replace a unit with one that has a lower working voltage because it can run real warm, red hot, or blow up in your face.

Earlier it was stated that although a constant hum (regardless of the channel selector position or adjustments) is most likely to be caused by a defective filter component, and that a hum that responds to a change in station or volume control setting is probably caused by a defective (although still functioning) capacitor, a filter-caused problem might produce effects other than audible hum. These problems include snaking or weaving, or other image distortion on the screen. Be sure to looks for any abnormal displays on the screen.

Buzz

By buzz, I mean a sound characteristically peculiar to TVs (except the very ancient ones that have a split-sound system), and originating in the TV circuitry incident to normal operation, but caused by one or more misadjustments. This buzz is similar to the hum described earlier, and is of the same pitch (60 hertz), except for a raspy, buzzy quality, instead of the relatively smooth tone of the hum. Another type of buzz is not related to the electrical operation of the TV—a mechanical buzz. Although similar to the TV buzz, which is sometimes called *intercarrier buzz*, the mechanical type is more haphazard and random, and is easily identified. In either case, buzz is a defect and requires correction.

Before attempting to correct this defect, it is necessary to ascertain which type of buzz it is. The easiest to eliminate are the mechanical buzzes, such as the buzzing from the loudspeaker or anything loose in or near the cabinet. A dangling wire or even a piece of paper or cardboard near the set can easily be identified, located, and remedied by removal, tightening, etc. Loudspeaker buzz, although actually mechanical, is probably caused by a defect in the speaker assembly—a tear in the speaker cone, the rubbing of an out-of-round voice coil (usually warped by heat), or even a simple loosened mounting. Any of these types can be recognized by comparison (keeping in step) with the audio, as well as total disappearance when the audio momentarily pauses. Although a buzzing component is easily corrected, and a tear in the speaker cone can be judiciously repaired by application of quick drying cement to the tear (not patching with tape), a rubbing voice coil is almost never satisfactorily repairable. But it is quite within the ability of the average do-it-yourselfer to replace the whole speaker by following a few simple directions:

- Disconnect the speaker. This will usually be a pair of wires connected by push-on clips, automotive style. Sometimes it is a two-pin plug. Others will be soldered into the circuit. Dismount the speaker from the cabinet to gain access to the cone (speaker front).
- Carefully press cone near its center so that it moves into the assembly, at the same time listening and feeling for any scraping or binding. There should be absolutely none. If this is the case, the speaker is not at fault.
- Obtain a duplicate speaker. Any parts supply house will identify the speaker and furnish a replacement. Of course, the physical dimensions (4" by 6", 5" by 7", 4" by 10", 6" by 9", etc.) should be the same, otherwise there will be a fitting problem in the cabinet. Be sure that the impedance rating is the same (4 Ω, 8 Ω, 16 Ω, etc.).
- Unsolder the old cable, if used, from the defective speaker and solder it to the lugs or terminals of the new speaker, taking care not to scorch or burn the cone in the vicinity.

If there is any question about the characteristics of the old speaker, a reference to the data package for the TV set (the parts list or the schematic diagram) will sufficiently identify the speaker to enable the replacement parts house to find a replacement.

The preceding examples were either mechanical types of buzz issuing from the offending part, or generated by the speaker in step with the sound. The following types are nonmechanical, but they emanate from the speaker. What is most characteristic about these types of buzz is their pitch; they are all of 60-hertz pitch, although their timber is different. It is not essential for the purpose of this explanation to consider the technical reasons for this. Both hum (usually called *ac hum*) and buzz peculiar to the TV circuit operation are of the same basic frequency, which also is the frequency or repetition rate of the vertical sync. And although it was not intended that sync pulses reach the audio portion of the TV, and ultimately the speaker, such is sometimes the case, as detailed in the following section. If this occurs, the audio system responds as to any other sound and reproduces it in audible form.

As to the difference in sound between these, the musician would perhaps explain this by calling the ac hum a pure (single) tone, and the buzz a complex tone, one having a mixture of many overtones. As in case of hum, some characteristics of buzz help you locate its origin within the TV set. These are covered in the following paragraphs.

Warm-up buzz

During the first seconds (up to a minute or two) of the warm-up period, a buzz does not necessarily indicate any malfunction; a stabilization of some of the automatic gain circuitry usually reduces this buzz to an inaudible level. No action is required in this case.

Constant buzz

If the buzz remains for any length of time or is a constant annoyance, simple adjustments might remove it. The checks for the specific cause should be made. Often, it is necessary to set the fine-tuning control, not for the best sound consistent with a good picture, but for the optimum sound with good picture and minimum buzz. Once this control is adjusted, it should hold for most stations without individual adjustments. If this does not reduce the buzz to inaudibility, overloading might be responsible.

Intercarrier buzz

This buzz is inherent and characteristic of almost all TVs, except the ancient ones known as the *split-sound types*. A great many TVs have a rear-chassis adjustment (usually a slotted-shaft type) marked simply *buzz*. The control is rotated very gradually until the buzz is at a minimum. This should be done with a station tuned in while listening to the sound (music or speech) accompanying the program.

Some TVs have a variable coil (quadrature or discriminator), which is adjusted to eliminate buzz or tunable hum in the sound. A manufacturer's schematic and parts layout will help you to identify the correct coil. (Do not attempt to adjust the sound coils unless you know exactly which one it is.) The sound coils are often located next to the sound output transistor, or IC output (Fig. 7-16). Usually, the tallest shield coil is the interstage, and the short one is the discriminator or quadrature coil.

After locating the correct coil, check for the type of adjustment tool needed. Some coil slugs have screwdriver slots, and others require a hexagon-shaped alignment tool. Choose a strong TV signal and tune it in the best picture, color, and sound possible. Remember, a touch-up of this coil is all that is needed. A slight turn clockwise or counterclockwise might be sufficient. Simply adjust the coil until all or most of the buzz has disappeared. Recheck the picture and sound reception from several different stations.

7-16 Proper adjustment of the discriminator coil can eliminate buzz and distorted sound.

Multiple pictures

I refer here to the existence of two or more duplicate pictures, one on top of the other. Usually each picture is not very stable and is only a fraction of the full screen height. The fault here is unmistakable: the vertical oscillator circuit is operating out of frequency. There are three possible causes, two are reasonably likely, and the third is very rare indeed.

Vertical hold misadjusted

Whether this control is an operating control on the front panel or is semiaccessible one behind a little door, it usually has sufficient adjustment latitude to cause oscillator operation at a fraction of its normal frequency or well above it. In other words, it is possible to produce such a malfunction by rotating the control to its extremes. The first step to correct a case of multiple pictures is to adjust the vertical hold control first in one direction, then in the other. The picture will roll up or down, and after a while should jump from three to two and finally to one picture across the screen. Careful adjustment is required after a single picture is obtained to leave the control at the optimum position. This is checked by switching stations and observing whether or not the picture rolls (Fig. 7-17).

Horizontal multiple pictures

If the horizontal control knob will not stop the picture from rolling or moving sideways, suspect that a horizontal oscillator circuit is defective (Fig. 7-18). Look for a defective horizontal oscillator transistor, IC, or poor filtering of the horizontal circuits. Within the transistor circuits, the horizontal sync can be scoped, indicating a poor or missing horizontal sync. Dried-up decoupling capacitors filtering the horizontal oscillator circuits can cause poor horizontal lock in.

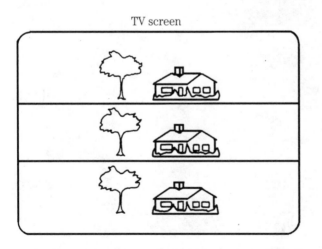

TV screen

7-17 If the picture rolls upward or downward, readjust vertical control.

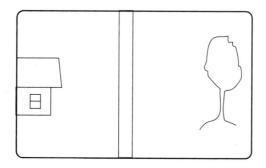

7-18 If the picture is split, suspect that there is no horizontal sync control.

Horizontal drifting can occur if the chassis is normal for a few seconds, then the picture goes into horizontal lines. The horizontal hold control might straighten the picture for a few minutes until it starts tearing in another direction. Check each bypass and filter capacitor in the horizontal hold circuits. Sometimes spraying the area with coolant can help you to locate the defective component.

Defective component

A defective capacitor or resistor (changed value) in the vertical oscillator circuit will cause something other than one picture to appear on the screen. In rare cases, the vertical hold control can compensate for this change, but this is most unreliable. If the control has to be moved to an extreme position, a component is at fault and the cure (at best) will be very temporary.

Other causes

What has been said about multiple pictures or scenes in the vertical direction, applies to some degree also to the horizontal direction; however, in the case of horizontal off-frequency operation, either caused by misadjustment or (more likely) a defect, the whole image might be much less stable, going through any number of variations (Fig. 2-47), during which no recognizable picture can be displayed.

What is common to both types is that there is a tendency for the scene to lock or stabilize at a submultiple of the correct repetition rate. Thus, the vertical repetition rate is 60 Hz, and the transmitted sync pulses normally ensure locking at exactly this rate, provided that the circuit is initially set as close as practical to 60 Hz. In other words, the transmitted sync information can and will correct minor deviations from exact frequency. If this deviation is excessive, however, sync correction is no longer effective and the picture rolls or tears. Apparent locking or sync at a frequency that produces two or three pictures across the screen is what might be called a *technical quirk* (the explanation is beyond the scope of this book) occurring at exactly half or a third or sometimes even a fourth of the correct frequency. If a malfunction causes the vertical sweep to operate at approximately 30 Hz, the sync will more or less hold, but two pictures will now be seen. Similarly, three images would appear if the frequency is around 20 Hz. Sync is less effective at each lower submultiple.

In case of the horizontal direction, the basic frequency is much higher (15,750 Hz). The number of the heavy stripes (Fig. 2-47) corresponds to the frequency on which the sweep circuit is operating.

Multiple pictures in the horizontal section might result from improper adjustment of the horizontal hold control. For example, after a loss of horizontal sync, the H-hold control is turned the wrong way. Instead of the picture straightening up or locking in horizontally, all you see is lines and multiple pictures in a horizontal plane.

This condition can occur in small black-and-white or color portable TVs, where a single coil adjustment is used as the horizontal hold control. These controls have no stops in either direction and can be turned either way until the adjustment slug drops out (Fig. 7-19). If the coil slug or core has fallen out of the coil at the back, restart the core from the back side. Often, the plastic shaft end has broken off, letting the core fall out. Replace the plastic shaft after threading the metallic core into the coil. Use the hot end of a soldering iron to make a flat end on the protruding plastic shaft so that it can not pull out.

Turn the shaft until the core is in about the middle of the coil. Connect the antenna and tune in a local TV station. Rotate the horizontal hold control while observing the TV screen to determine if the lines are spreading apart. Slowly adjust the control until the picture locks in. Next, quickly switch the channel selector off channel and switch it back on again to observe if the picture loses horizontal sync. Readjust the horizontal hold control until all stations remain locked in with the set hot or just turned on.

Horizontal hold control

7-19 Locate the horizontal hold control on the rear chassis in the older TV chassis to remove horizontal bars in the picture.

The preceding explanation of multiple picture causes is intended to aid the do-it-yourselfer to analyze and locate the defect in such cases. Keep in mind that, by design, the sweep frequencies are pretty close to the exact values, and the adjustments are fine adjustments made to compensate for minor deviations caused by normal wear and tear. It should be obvious to the troubleshooter that a severe or extreme condition cannot be adjusted back to normal.

Various suggestions given here, particularly the adjustments for purposes of correction, are intended for, and will be effective only in case of, wear and tear deterioration or gross misadjustment through carelessness. In case of a component failure, whether a transistor or passive part (such as a resistor or capacitor), no amount of adjustment is better than a permanent cure.

Even in those cases where setting the control to or near one of its extreme positions of rotation brings back synchronization, the cure will only be temporary, because, as explained previously, the setting of the control is only approximate, to enable the sync pulse to take over. With the control at its extreme, no correction latitude remains, except in the opposite direction of rotation. In other words, the control will no longer serve its normal function of being adjustable up or down from its center position. With a slightly more degradation, no adjustment will be possible. Locating the defect and its correction is obviously the answer.

Vertical and horizontal lock-in

Today's TV might have a horizontal frequency control inside the chassis and no vertical or horizontal hold control to be adjusted by the operator. This is usually accomplished with processor countdown and crystal-controlled circuits. The circuits are so well regulated and controlled that no hand adjustments are needed. Of course, when the TV pictures roll, bounce, or slip sideways, internal sync circuits must be repaired.

Smoking chassis

Shorted or overloaded power transformers can emit a pungent smell. A arcing yoke winding might burn intermittently (Fig. 7-20). In the earlier solid-state chassis with an integrated high-voltage transformer, the enclosed molded diodes could arc over. The early tube flyback transformers had a tendency to arc and burn with no drive voltage on the horizontal output tube. Besides a burning smell, the leaky tripler unit might arc over between the plastic shell and the chassis.

Although transistors, resistors, and capacitors are physically small, they can short and burn holes in PC boards. Often, the board can be repaired by cutting out the charred section and rerouting the wiring with external hookup wire. After being struck by lightning, the chassis might begin to smoke in several sections. Usually, the chassis can be repaired by replacing the smoking component.

Line noise and filters

In connection with the problem of electrical noise or interference such as streaks, dashes, etc., some words about noise filters are in order. Many such devices or gadgets are advertised to cure any and all TV ills. Although they are not all completely useless,

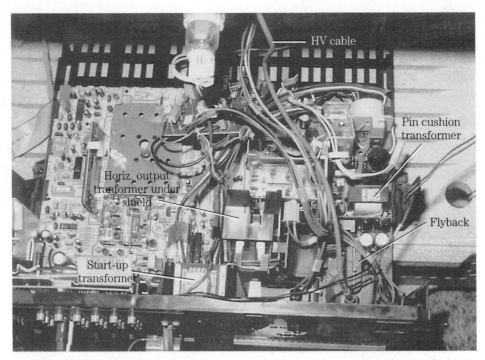

7-20 Check the following components for smoke, burn, or arcover in the TV chassis.

many of their claims are exaggerated. Before buying and installing one of these, you should know what you can reasonably expect.

Three sources of noise are associated with a TV set, whether visual (in the picture) or audible. Two of these are external, and the third is internal. Noise in a TV can result from a defective component; for example, an old resistor can sometimes become intermittent. In case of a defective component, a professional will have to be called on for analysis and repair. The same holds true for operating controls that become noisy with age. This includes contrast, volume, tone, fine tuning, and the channel selector.

Line noise includes all types of electrical noise generated by appliances that operate on ac. Although not all appliances are potential noise makers, some of are. Their electrical noise travels along the ac wiring in the house and enters the TV. By contrast, noise from utility poles equipment and devices outside the house seldom reach the TV this way. But this is nothing new, and radio and TV manufacturers have, almost without exception, provided simple noise filters inside the receiver for this purpose. Nevertheless, it is possible that an additional filter might be of some help if it is the proper kind and is properly installed.

It is very doubtful that the simple little gadget that looks like an ac plug on one end and an ac receptacle on the other can be much use. The fact that the built-in filter does not seem to help suggests one of two possibilities. Either the noise does not enter via ac line, or (if it does) the elimination of the noise must be accomplished at the offending appliance, (washing machine, oil burner, electric drill, etc.).

The third source of noise, and a very likely one, is the antenna. Of course, an old corroded antenna with poor lead-in connections is a potential source of noise. But what

I have in mind here is noise generated elsewhere and picked up by the antenna, along with station signals. Other than relocating the antenna and transmission line, nothing can be done here, except enlisting the aid of the utility company if the power lines, pole transformer, etc., prove to be the cause. In fact, for any, but the simplest type of local noise, the average TV owner is hardly in a position to move his or her antenna any appreciable distance — especially horizontally. A change in the antenna height might be helpful if it can be first found experimentally that a different height (higher or lower) proves noise free.

With regard to the transmission line, the problem is easier to solve. If it is apparent that the transmission line runs along a noisy structure or device or power line, either relocation or shielding should produce some improvement. Relocation is simpler and should be tried first. If this is not feasible or it does not help, substitute of a length of shielded transmission line, which is almost 100 percent certain to be effective.

Two simple types of shielded transmission lines are suitable for practically all TVs. One is unbalanced 75-Ω line, sometimes called coaxial cable. This contains an insulated center conductor and a metallic (braided) outer conductor, sometimes rubber covered. When installing this cable, the outer conductors must be thoroughly grounded by a solid connection to the TV chassis. In addition the TV must be equipped (nearly all are) for a 75-Ω transmission line. Otherwise, a little accessory known as a *matching transformer* must be connected between the coaxial line and the TV.

A second type of shielded transmission line is the 300-Ω twin-lead shielded line. This is the usual two-wire flat line with the addition of an external shield. Assuming that the TV requires a 300-Ω antenna (most sets do), the connections are the same as with the unshielded transmission line, except that the external shield must be tightly connected to the TV chassis at one end and to the TV mast at the other. There's usually a slight signal loss because of this outer braid or shielding, but this is not significant, provided that the antenna proper is fairly good.

In extremely noisy areas, use a shielded coaxial lead-in and place the antenna as high and as far away as possible. If a noisy power line is radiating noise across the reception path of the TV antenna, place the antenna on a higher tower. The height of the noisy power line and its proximity to the TV antenna should be considered when deciding if the additional cost of relocating the antenna is warranted. In some areas, it might be impossible to eliminate all noise from the low VHF TV channels. Always use shielded cable to keep the power-line noise from being picked up by the lead-in wire.

The notion that an indoor antenna is immune to outdoor noise is completely false. No compact, simple, abbreviated antenna can improve reception over a good outdoor antenna. In fact, the window improvements can seldom approach the performance of even the simplest outdoor antenna that has been properly installed. Except for locations with extremely high signal strength, these window gadgets are fairly useless. And as to devices that use radar principles and convert one's house-wiring into one giant all-direction antenna, the best that can be said for them is that they will work where any scrap of wire will work; they will not be much use where a normal antenna is required.

Spike protectors

The spike protector can protect the present-day TV set, stereo, word processors, computers, test equipment, and telephone systems from transient spikes from static electricity, motors, lightning, etc. Some spike protectors also suppress radio-frequency

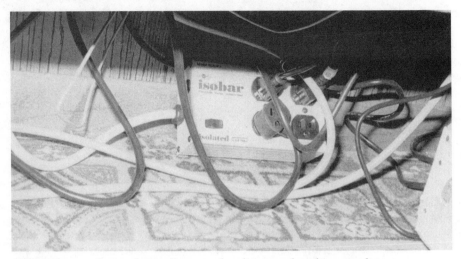

7-21 Voltage spikes and transients can be eliminated at the ac outlet.

interference. The spike protector cleans up the ac power for computers and helps prevent data loss, false printouts, monitor distortion, and computer component damage.

Some units are very small, with a three-prong plug that plugs into the ac power receptacle, then the TV is plugged into it. The large units might have four or six outlets with indicator lights (Fig. 7-21). The clamping response time might be within 5 nanoseconds and suppresses up to 220 joule spikes at current levels up to 13,000 A. Often, the unit will handle up to 15 amps of load power.

If your new TV keeps breaking down every few months, suspect that transient spikes are destroying IC or critical processors. In today's busy world, many power motors and high-voltage power lines contain transient spikes that might destroy the small control processor. If you live in an industrial neighborhood, place a spike block between TV and ac outlet.

Power-line surge protectors

In some of the latest TVs and microwave ovens, a surge protector prevents extensive damage to the ac components. A high-voltage line surge or storm line damage might place more than 120 volts on the TV. The surge protector might prevent damage from a lightning storm. A TV can be protected by a surge protector if lightning strikes a tree or power pole two blocks away.

The surge protector might be included in a spike block, separate unit, or soldered across the ac line in the TV chassis (Fig. 7-22). These surge protectors look like ceramic disk capacitors, except that if a higher voltage surge is on the power line, they will short and destroy the ac fuse. Actually, the protector takes the surge of voltage, so to speak. These single-surge protectors can be added to the chassis of any unit that is powered by the ac power line.

FM traps

Unwanted interference from strong local FM stations can be eliminated with one or more FM traps. The broadcast signal from a strong local FM station will sometimes modulate (ride on top) the TV station signal and cause FM oscillating lines or curved

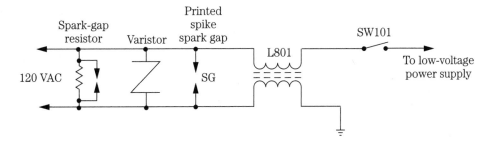

7-22 A disc spike varistor, spark gap, and printed circuit board spark gaps at the ac input can protect the TV from high-voltage surges.

patterns across the picture. The FM signal might be picked up by the TV antenna as a harmonic of the TV broadcast frequency.

To eliminate the interference, place an FM trap across the antenna terminals at the rear of the TV. If the FM signal is still seen in the picture or heard in the sound, a second FM trap must be placed at the TV antenna terminals. This means that the FM trap must go high up the antenna mast. Change all the flat lead-in wire between the two FM traps to a coaxial shielded cable to help eliminate FM or unwanted noise from being received by the wire (Fig. 7-23). If a booster antenna system is used, be sure that the top side of the booster has an adjustable FM trap to help eliminate unwanted FM signals.

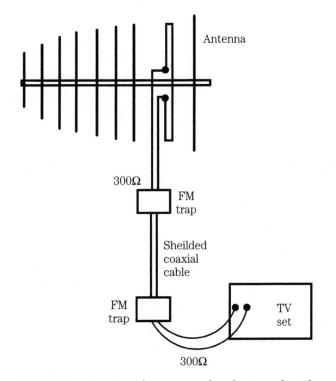

7-23 FM station interference can be eliminated with shielded cable, and an FM trap at the antenna and input terminals of the TV set.

8
CHAPTER

Troubleshooting
catastrophic failures

The various adjustments and corrections covered so far have been primarily of the preventive-corrective maintenance type. They are intended to keep a TV set in normal operating condition, as well as to correct deviations caused by aging, wear, deterioration, etc. Our concern now involves catastrophic failure, where the TV set ceases to operate either partially or totally.

Completely dead receiver

As obvious as this might seem, it is sometimes overlooked. If there is absolutely no evidence that the power is on (pilot light on, etc.), a few simple tests will provide the answer. They should be made before going further.

- Check for a loose plug, or a disconnected plug at the wall outlet.
- Check for a defective wall outlet by plugging in a lamp or another appliance.
- Check for a defective fuse or an open circuit breaker. In the earlier sets, a power-line fuse is located on the rear apron of the TV chassis (the metal strip at the bottom of the set). More recent receivers are likely to have a resettable circuit breaker that can be reset or closed by a simple push. This requires a few seconds for cooling off after the set goes off. If the circuit breaker opens again immediately after resetting (or, for that matter, should the replacement fuse blow similarly), there is a defect, such as a severe overload or short circuit. Further troubleshooting is required.
- Check the on/off switch on the TV set. It might be defective.
- Check for a loose or not-completely-inserted interlock plug of the ac cord. If the pilot light and picture seem to operate intermittently, suspect poor ac plug connections (Fig. 8-1). In older models, the interlock plug is disengaged if the back cover is removed for customer protection. The ac contacts might become worn or corroded. Lightning might cause a no-picture or

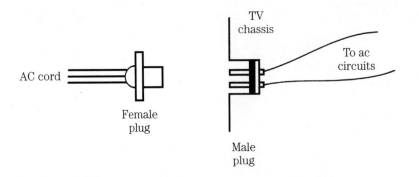

8-1 Intermittent TV operation can be caused by a defective interlock plug and socket.

intermittent operation. In modern TVs, the ac cord is soldered into the ac circuit. Replace the female chassis plug, if it is excessively burned or has poor contacts.

If the answer to the first two checks is negative for transformer-type sets, the fault lies either in a defective on/off switch or a defective power transformer. In either case, the repair is not for a beginner. The switch is almost always part of a control assembly (volume, tone, etc.) and requires professional attention. A transformer replacement is even more out of the reach of an amateur. Do not attempt either. Fortunately, both of these possibilities are rather improbable.

Isolation transformer

The isolation transformer is used by electronic technicians to prevent blowing fuses and chassis damage when connecting test equipment to the ac TV chassis. The TV that works directly from the power line, without power transformer, might have a hot chassis and should be protected with the isolation transformer (Fig. 8-2). Sometimes, when applying the ac-operated test instrument, something has to give. The common ac ground of the test instrument and common hot ground of TV chassis might blow more than fuses in the ac TV chassis.

The ac TV chassis is plugged into the isolation transformer and the transformer is plugged into the power receptacle. Besides isolating the power line, the transformer might have adjustable taps to raise or lower the power line. Raising and lowering the power line in the ac TV chassis might be required to service chassis or high-voltage shutdown. Slowly raising the power line might prevent damaging another horizontal output transistor. An intermittent problem might be located by raising or lowering power-line voltage.

Raster only

With only a white screen, the horizontal, vertical, picture tube, low-voltage power supply, and high-voltage circuits are normal. If there is no sound, the trouble can be between the antenna and the IF-video take-off circuits (Fig. 8-3). If the audio is normal, the trouble likely occurs in the video output circuits. To eliminate the

8-2 Use a variable isolation transformer when working on any TV to protect test equipment and fuses.

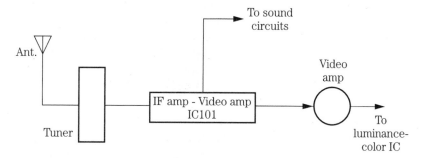

8-3 Suspect the tuner, IF stages, or video amplifier if the TV shows only a white raster.

tuner, connect a tuner subber to the IF cable. In fact, the IF cable will plug into most substitute tuners. The video signal can be scoped from the IF circuits to the luminance and color IC circuits with the oscilloscope.

No sound/normal picture

The no-sound symptom can be caused by a defective component from the IF audio take-off at SIF, to the speaker. Dead sound usually is caused by a problem in the speaker, speaker coupling capacitor, output transistors, or IC. Rotate the volume control up and down; notice if a rushing or scratching sound is heard. Place a screwdriver blade on the center terminal of the volume control with the volume control wide open. A hum and rushing noise in the speaker indicates that the output circuits are normal.

If there is still no audio or hum, check the audio output circuits. Clip a speaker across the suspected speaker to determine if speaker voice coil is open. The speaker can be used as an audio signal tracer with a small electrolytic capacitor to check the input and output audio signal of the output transistor and IC (Fig. 8-4). With normal volume, the volume control will be real low at the input and very high at the output. The volume should increase as each stage is signal traced from the speaker to the input stage.

Weak sound can be caused by a defective audio stage, transistor, IC, electrolytic coupling capacitor, speaker, or by an improper low-voltage source. The weak stage can be signal traced with an external audio amplifier. Shunt the suspected capacitor with another and check for sound. Check the signal from the volume control going into the sound input terminal of the output IC. Then check the sound out at the IC output terminal.

Distorted sound can be caused by a defective speaker, leaky coupling capacitor, leaky output transistors, burned bias resistors, or a leaky output IC (Fig. 8-5). Again, signal trace the audio in at the base terminal of the driver or AF transistors. Likewise, do the same on the input and output terminals of audio output IC. Check the distorted signal up to the sound output transistors. Excessive audio distortion can sometimes be found in the audio output circuits. Take an in-circuit test of transistors, for leakage, open, or normal conditions. Test each bias resistor when the sound output transistors are out of the circuit. Sometimes a leaky transistor will burn or destroy the bias resistor.

An important caution must be made in connection with the sound section of a TV, which also pertains to the handling of the speaker. Because most speakers in TVs have quick-disconnect clips or pin-and-jack connections (for convenience in removing the chassis from the cabinet), it is quite possible to deliberately or accidentally pull the speaker leads off, leaving the speaker electrically disconnected from the receiver. This is potentially damaging to the output transistors and the IC.

Sound IF circuits

The 4.5-MHz IF signal is taken from the video-IF IC or transistor. A TV signal-processor IC might contain a limiter, FM detector, attenuator, and audio preamp circuits. The 4.5-MHz sound discriminator coil is taken from the FM detector. A mushy-distorted sound might result from misadjustment of the 4.5-MHz discriminator coil or transformer (Fig. 8-6). Sometimes too much moisture will change the coil

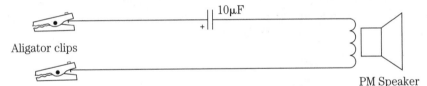

8-4 No sound/normal picture indicates that problems are in the sound circuits. Signal trace audio output circuits with speaker and a capacitor.

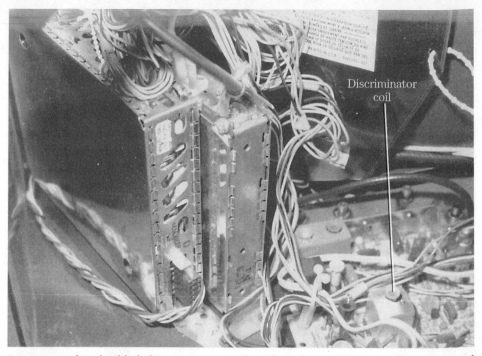

8-5 Locate the shielded discriminator coil and readjust if the sound is distorted and mushy.

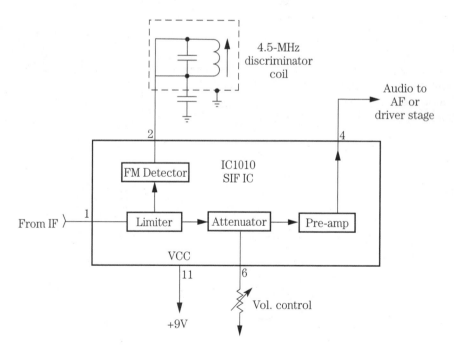

8-6 The IF signal is fed into the SIF IC (IC1010) and the discriminator coil is at terminal 2 for adjustment.

frequency. The limiter stage removes any AM modulation while leaving the FM sound information. The 4.5-MHz discriminator coil tunes in the intermediate frequency.

The volume control provides voltage to the attenuator circuit controlling the amount of sound passed on to the audio preamp circuit. The audio output is fed from pin 2 to a transistor or IC to drive the small speaker. Some audio circuits do not use the attenuator or preamp circuits inside the IC. The audio output circuits can be either transistor-operated or a single IC (Fig. 8-7). Often, three transistors are used in the transistorized output circuits.

No sound, but picture OK

Often, the no-sound symptom with normal picture indicates problems within the sound stages. In present TV chassis, the horizontal stages must perform before sound will come up to normal operation. These sound problems usually occur in the output stages. The sound stages might consist of only one large IC (Fig. 8-8).

The early solid-state sound circuits consisted of transistors, and then integrated components were added. In some chassis, one IC might contain the audio IF amplifiers, discriminator, preamp audio, and audio output circuits. Today, the audio IF, FM detector, and audio amp might be included in one large IC (Fig. 8-9). The sound output is amplified by either transistors or another audio output IC.

The audio might be signal traced with another audio amplifier or signal traced from signal point A. If you find no audio at point A, suspect a defective IC or improper supply voltages. If you find sound at the input of the audio output transistor and not at the speaker, check for a defective transistor or speaker. Use a signal tracer and check the audio signal right up to the speaker terminals.

Snow

This picture defect, snow, might vary from good sound/snowy picture to fairly good sound/very faint picture, or no picture at all. In addition, the snow might appear at random, suggesting (improperly so) that the fault lies outside the set. Actually, this is not so. *Snow* is a name given to a picture that seems to be broken up into

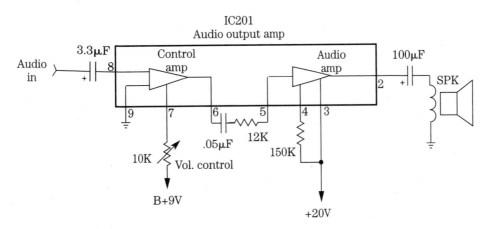

8-7 The audio-output IC amplifier controls the audio to the speaker.

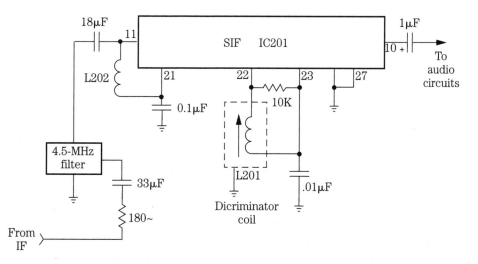

8-8 Adjust L201 if the sound is distorted and weak.

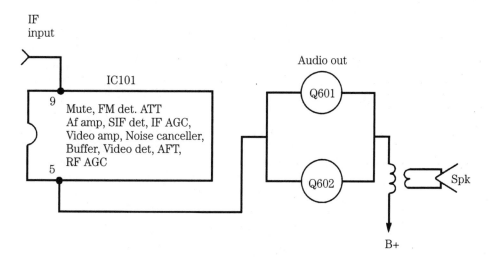

8-9 A block diagram of an IF/SIF/detector/video/AFT IC with sound fed to two separate audio transistors with no picture.

small, wavering flecks or pieces. They do not move down the screen, although they are somewhat reminiscent of snow-filled air.

Regardless of which of these snow problems afflict a particular TV, it is distinctly different (in fact, opposite) from the weak-picture cases described earlier. In the vast majority of cases, the presence of snow is evidence that the TV is working hard to produce a picture from a poor signal. This means that either no signal is received by the TV, or that the received signal cannot get through because of a break in the functional continuity in the set. The procedure for isolating the fault and applying the appropriate remedy is simple.

Examine the antenna lead-in system. An old or otherwise poor antenna system will not necessarily cause a sudden appearance of snow. But a break in one or both conductors of the lead-in at the point of connection to the antenna, at the point or points of fastening on the way down, at the entrance through a window or wall, or at the connection to the antenna terminals on the rear of the set is a likely cause for snow on an otherwise-normal TV. To check for a suspected broken lead-in, a substitute antenna, rabbit ears, an indoor wire antenna, or even a length of wire connected in place of a suspected antenna will be sufficient to verify whether or not the antenna is at fault. Of course, make allowance for the mediocre performance of the substitute antenna. But if the picture improves and the snow effect is diminished, the antenna's transmission line is probably at fault, so you must check for a break in the wires.

A snowy picture might be caused by improper orientation of the antenna or from missing antenna elements. Check and see if the wind has turned the antenna away from the desired broadcast signal. Doublecheck the antenna's direction, in respect to the control-box indicator, if a rotor is being used to orient the antenna in fringe areas. If snow is absent from the picture on some channels, but is visible in the picture of a lower numbered channel, suspect that several of the back elements of the antenna are missing. The signal could be cut in half if just the back element rod is missing.

Snowy tuner

A turret tuner can be snowy with a leaky or open RF transistor and dirty tuner switching contacts. Clean the tuner contacts with tuner lube. Spray into each switch contact. Be sure that the outside antenna, lead in, and connections at the TV do not produce a snowy picture.

A defective varactor tuner in the latest TVs can cause a snowy picture, drift off channel, or appear dead on several channels (Fig. 8-10). Often, these tuners cannot readily be repaired and must be sent to the factory tuner repair depot or order out a new one. Varactor tuners that are in warranty must be returned to the manufacturer for replacement. The latest tuners can be soldered to the PC wiring and can be removed with solder-wick and iron.

Although the tuner is a most likely suspect for this defect, it is assumed that tuner is operating normally because, as mentioned earlier, tuner abnormalities evidence themselves in other ways and their repairs will automatically preclude them as a source of snow. Nonetheless, and especially in cases of minor deterioration in tuner operation, which many TV owners are willing to tolerate for a while, imperfect contact in one or more of the tuner switch contacts will also cause a snow effect on the screen. Usually, in the early stages of tuner wear, it is necessary to jiggle the tuner channel selector to obtain a better contact and eliminate the snow. Incidentally, this is an obvious hint that the tuner needs maintenance—cleaning at least.

Snowy picture

Do not overlook the possibility of lightning damage if the picture is snowy. Sometimes lightning might strike a large tree near the power line and cause only minimal damage to a TV one block away. Inspect the antenna terminals for black or burned spots. Sometimes the one antenna wire might be blown off. The center wire of the coaxial cable might be fused to the connector or to the outside shield.

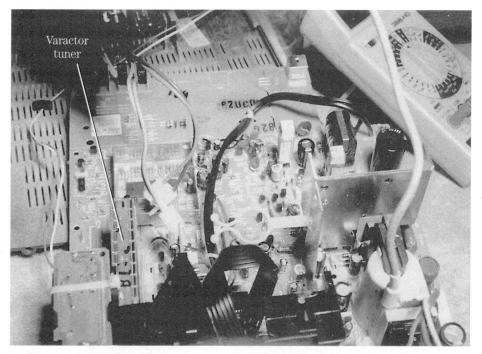

8-10 The varactor tuner might drift off channel, and become snowy and leave only a snowy raster with no picture.

Check the balun coils located on the tuner. Notice if the small coupling disc capacitors are burned. These two capacitors isolate the antenna lead-in from the antenna input to prevent damage within the TV chassis (Fig. 8-11). Sometimes the coupling capacitors are burned or blown away with damaged balun coils. You might find only one or two ends of the coil burned off. Simply soldering back the burned leads might cure the snowy picture. Continuity tests with the ohmmeter can easily help you to locate the open balun.

Sound OK, but no picture

Your TV might have a totally blank screen with normal sound points, or trouble in the video portion of the TV, including the picture information and horizontal sweep circuits (if there's no raster), but not the vertical or sync circuits. The horizontal stages are involved because they are responsible for the presence or absence of any light (raster or picture) on the screen.

The troubleshooting procedure is based on some logical assumptions. First, the picture tube is not burned out (Fig. 8-12). In a transformer-type set, a burned-out picture tube is identified like any other burned-out tube—absence of any light near the base of the tube. Of course, it is assumed that the picture tube socket is properly seated on the tube base. Once in a long while, it might work loose or simply lose proper contact. Second, the tube did not become defective since the last time it was on; it usually doesn't happen that way, certainly not suddenly.

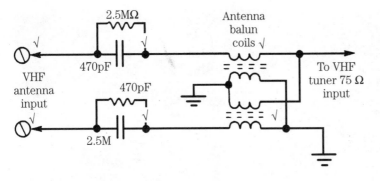

8-11 Check for burned antenna input capacitors or balun coils if you suspect that the TV has lightning damage.

8-12 Check for a lit heater or filament at the neck of picture tube. Wiggle the CRT socket to see if tube will light, indicating that the tube socket is defective or that the element pins are dirty.

The heater or filament might not light up with a defective socket, poor wire connections, or PC board connections. Sometimes slightly moving the CRT socket might turn up a defective socket. Some of these sockets are enclosed, and others have spark gaps on the picture tube elements.

In the early TV chassis, the thin wire from the picture tube heaters would have a poor contact and cause intermittent tube filaments. The TV set might operate for hours and the screen go black. Simply soldering the pin socket on CRT solved the intermittent heater problem.

The CRT socket might have bent or burned socket pins and caused the heater to not light up. Check where the filament pins solder directly to the tube socket. Sometimes soldering these pins solves the intermittent picture. Check where the chassis filament wires connect to the picture tube socket. Also, check the filament wire connection on PC board or chassis for poorly soldered connections. Take a continuity ohmmeter test across the filament pins of the socket for low-resistance measurement. No reading or a high reading indicates that the connections are open or poorly soldered (Fig. 8-13).

As stated earlier, some TVs use an internal fuse, usually inside the high-voltage cage, to protect only the horizontal output portion of the TV. This is not the main fuse, which protects the whole TV. When this internal fuse blows, the whole horizontal output circuit, including the high-voltage system, is interrupted, resulting in no light on the screen. Replacement of the fuse is relatively simple, even in those cases when the fuse is soldered into position by pigtails (Fig. 8-14).

Visual inspection must be very carefully made as the metal conductor inside the glass fuse is very thin; therefore, an open is not too obvious. If the fuse is a snap-in type, the procedure is quite obvious. The fuse should be of the same physical length to fit into the fuse-holder clips. Its electrical value is very critical. If the fuse is of the soldered-in type, a slightly different replacement procedure is called for. All radio-repair supply houses sell replacement fuse assemblies.

Fusible or isolation resistors

In the early black-and-white and TVs, fusible resistors were used in series with the ac power line, instead of with the fuses. The fusible or low-resistance isolation

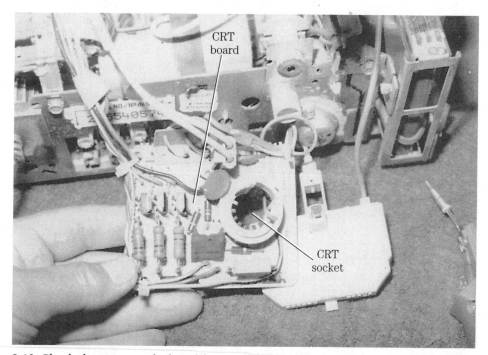

8-13 Check the picture-tube board for poorly soldered connections or dirty tube pins.

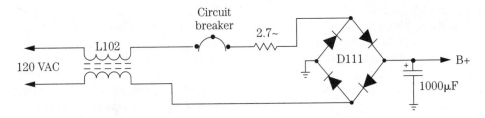

8-14 The circuit breaker will open with an overload in the low-voltage power supply or chassis circuits.

resistors provide time lag and get red hot before opening. The fusible resistor might be off by 2.8, 4.8, or 7.5 Ω. These fusible resistors were used in early high-filament or heater-type tube series circuits.

A small resistor might be used in the ac-powered TV. When a leaky or shorted component is located in the low-voltage or HV circuits, the low-wattage resistor opens to protect the other components (Fig. 8-15). The low-resistance resistor might be used in series with other fuses. Sometimes, when lightning or a component shorts within the power supply or horizontal output circuit, the fuse blows and the low-resistance resistor opens. Always check for a shorted low-voltage diode, regulator, or horizontal output transistor if a low-voltage resistor is open.

Dim picture-poor width

If the operation is just a little below normal, that is, if the picture could be a little brighter and have a little more contrast, or if the picture does not reach the edges, the fault is probably in the low-voltage source or horizontal output circuits. Check the voltage source supplied to the horizontal output transistor. If it is low, suspect a low-voltage component in the power supply. Readjust the B+ or HV control, if one is located in the low-voltage circuits.

Check for the raw dc voltage applied to the voltage-regulator circuits. This voltage on the line-operated TV can be from 150 to 165 volts dc. Suspect a main filter capacitor if this voltage is low. Shunt a new one across the main filter capacitor and notice if the voltage raises and the sides of raster return to normal width. Check for damaged isolation resistors with an increase in resistance from the bridge or half-wave rectifiers.

Next, check the voltage regulators if the correct voltage is applied from the main filter capacitor. A defective line-regulator or burned bias resistors can change the output voltage. Check for leaky zener diodes in the regulator voltage sources.

Check the horizontal output circuits if the correct dc voltage is on the collector terminal of the output transistor. Test the H.O.T. (horizontal output transistor) in the circuit. Measure the drive voltage at the base terminal and take a scope waveform test. Defective safety or hold down capacitors can cause poor width or excessive high voltage. Check the horizontal drive transistor, driver transformer, and voltage applied to the driver transistor. Doublecheck the high voltage at the picture tube with the HV probe.

Low ac line voltage

Low or high ac line voltage might cause problems within the TV. Low line voltage might cause the sides of the raster to pull in and make components work harder. Excessively high line voltage might shorten the life of the components or cause high-voltage or chassis shutdown. The correct line voltage should be about 120 volts ac. In country power lines, if the TV is the last on the power-line transformer, low voltage might occur (Fig. 8-16).

Suspect the high power-line voltage if the horizontal output or low-voltage power supply components repeatedly break down. Check the power-line voltage with the

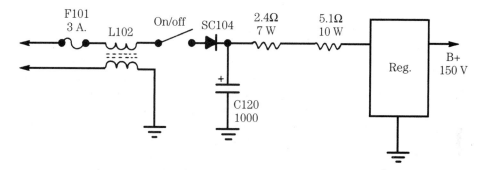

8-15 If F101 and the 2.4- or 5.1-W resistors open, suspect that the regulator or horizontal output transistor are defective.

8-16 Check the outlet voltage where the TV is connected.

DMM. If the power-line voltage is over 125 volts, notify the power company. Most power-line sources will place a standard voltage tester to be monitored over a 3- or 5-day period. Suspect the high power-line voltage if the chassis shuts down, kicks out circuit breakers, or blows the fuse for no apparent reason.

Picture dim and brightness/contrast ineffective

Another case of improper brightness or contrast control is when the picture size is normal. Unlike the previous example (where the brightness and contrast controls are functioning, but at their maximum settings), there is still not sufficient picture brightness. This case identifies a condition in which the brightness and contrast controls are not functioning properly. Sometimes increasing these settings (advancing the controls) will produce a photographic negative effect—blacks are white and vice versa. More often, the gradual variation from dark to light to bright, as is the case with the average TV screen, will not be obtainable, except perhaps at very low (dim) settings of the brightness and contrast controls. At all other positions, as when attempting to get a normal picture, the light and dark areas appear flat, muddy, and very dull. Sometimes such a picture suggests looking through a very dirty window or through a gray filter at a normal scene. Any of these symptoms, alone or in combination, suggest a defective picture tube (of course, it is assumed that the problem is not the one just described). The electron emission has deteriorated to such a low fraction of the normal amount that internal adjustments will no longer help. The only permanent remedy is the replacement of the picture tube, which is a job for the professional servicer.

Before deciding to have a new picture tube installed, determine if the cost of the repair is warranted. Replacing the picture tube is not prudent because you will soon have to replace the filter capacitors or the flyback transformer, which can be rather expensive. The expected life of a TV is from 10 to 12 years.

If you're shopping for a replacement picture tube (plus installation), some money can be saved by buying a rebuilt tube. In fact, if no specific instructions are given to the contrary, the replacement tube might well be a rebuilt one. This is quite satisfactory, both ethically and technically, provided that it is a tube rebuilt by one of the standard tube manufacturers. Brand new replacement tubes are still available, but their use is the exception, not the rule. A new CRT offers absolutely no advantage over the standard rebuilt tube.

When it has been established that a dim picture is caused to a weak picture tube, the life of such a tube can be extended, often for a number of months, by a simple technique commonly called *rejuvenation*. What it amounts to is a rejuvenation of the electron-producing element of the tube. It can be restored, for a while at least, by the application of a higher-than-normal voltage, raising the temperature of the electron-emitting surface on the tube element.

Most professional servicers use a one-shot remedy by applying an overvoltage for a short while, thus reactivating the electron emitter. After this, the tube reverts to its normal operating voltage. You might also install a booster device, which is attached to the tube and left there for the remaining life of the tube. It is simple, less expensive, effective, and most important, it might be less likely to shorten the tube

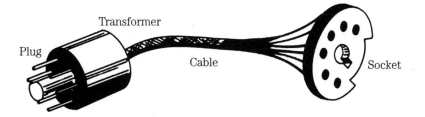

8-17 A typical picture tube booster with socket. The socket goes over the CRT base.

life. Because a rejuvenation overvoltage is applied for only a relatively short time, it is necessarily more drastic and might have a delayed effect on the remaining life of the tube. By contrast, the boost, which continuously operates with the tube, applies a much more modest overvoltage and is, therefore, less detrimental to the remaining tube life.

Figure 8-17 is a sketch of a common picture tube booster. At one end is a socket (or section of one, as used on most TV sets) that fits over the picture tube base, exactly like the original one. At the other end is a plug that contains a small transformer. Connection to the TV set is simple.

Switch off the set and carefully pull the socket from the picture tube base. Mate and connect this socket with the base plug on the booster. Connect the socket on the other end of the booster cable to the base of the picture tube. The booster can be left hanging.

One important reminder: There are two different types of boosters for sale in radio supply stores and they are not interchangeable—even if they do look alike. One is exclusively for parallel (transformer-type) tube hookups and the other is exclusively for series tube hookups (transformerless). Be absolutely sure which type of TV set yours is and purchase the correct type. Some types are for both filament arrangements and by virtue of a movable wire connection.

Heavy black-and-white bars

This symptom, heavy bars, looks somewhat like Fig. 8-18A. There might be either one dark and one light horizontal bar, each covering approximately 50 percent of the screen height, or, as in Fig. 8-18B, there might be three bars, one wide and two narrow. Often these bars will slowly drift up or down the screen. To locate the cause of this malfunction, more than one step is usually required.

Understanding the problem makes the solution much easier. The vast majority of such bars are caused by what is commonly called *ac hum*, although the word *hum* usually refers to the audible manifestation of the unwanted presence of ac in a circuit.

Hum bars might be caused from a poor filter capacitor. Heavy lines can be produced by a nearby high power line. If the bars remain in the picture with the TV antenna disconnected from the set, the problem is in the TV chassis. Interference or hum bars picked up by the antenna will not be seen after removing the transmission line.

One or two vertical hum bars in the raster can be caused by capacitors. Capacitors have a tendency to dry out and lose their capacitance, resulting in poor

A 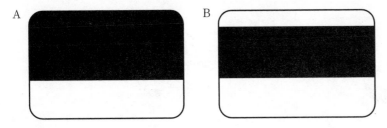 B

8-18 Ac hum can cause the raster to vary from black to white (A) or from white to black (B).

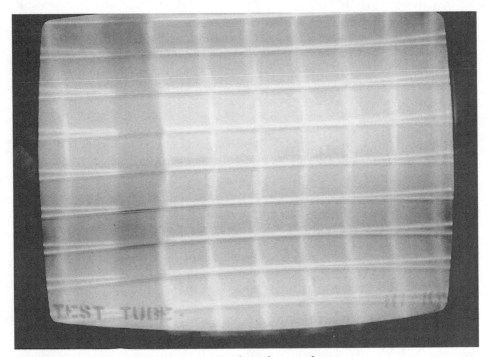

8-19 Defective filter capacitors can cause hum bars in the picture.

filtering of the low-voltage power supply (Fig. 8-19). A defective electrolytic capacitor can be located by shunting a good capacitor of the same value across the suspected one. Look for a white or black substance near the defective capacitor terminals. Check for correct polarity before turning on the set.

Very often, the fault will be in the power-supply portion of the picture tube circuits (that is, filter capacitor), but most likely, the picture tube itself has an internal short or a near short. Occasionally, even if rarely, it is possible to burn out the short with a tube rejuvenator. If a professional servicer offers to do this, he or she will probably stipulate that it must be done at your risk. This is not unreasonable and entails no loss to the owner because the tube is not serviceable and its trade-in value will not diminish—even if it is burned out in the attempt.

Sound bars

In connection with this phenomenon, this type of symptom stems several sources (overloading, AGC adjustments, etc.). The first is internal and is under the control of the user, the second is atmospheric, and virtually nothing can be done about it. Advancing either the contrast or the AGC adjustment beyond their normal settings, as described in Chapter 6 will give rise to various degrees of instability and even snaking.

In case of the external cause, little can be done. Normal TV reception is based on line-of-sight signal that travels from the transmitter to the receiver. The FCC-allocated station locations and power outputs is such that two stations of the same channel number are separated geographically so as not to interfere with each other. At infrequent times, however, and often incident to some radical weather change, signals travel not only directly (line of sight), but also via reflection from an electrically charged layer above the atmosphere, covering much greater distances than intended. Such a signal will play havoc with the local station of the same channel (because there are minute differences between the two stations, despite the fact that the same FCC standards apply to both), affecting sync, AGC, and other vital functions. The effect might differ from set to set and from station to station (the higher-frequency channels are usually not affected), but while it lasts, little can or should be done about it. Any attempt at correction through some adjustment will simply misadjust the set so that when conditions return to normal, the set will be out of correct adjustment.

One final possibility is overloading caused by the proximity of a powerful local station. Although relatively rare, this has happened in the past. In fact, an offending radio station actually distributed little devices—filters, of a sort—to residents in the vicinity of the powerful transmitter, to eliminate (or at least minimize) the problem. There are simple little signal reducers, which connect to the antenna terminals of the TV. Of course, the simplest remedy, even if it is not 100- percent effective, is to reduce the contrast control when the particular station is being viewed.

Sound bars are somewhat similar to the heavy black-and-white hum bars, except that they are much thinner (Fig. 8-20); more numerous (there might be a dozen or more); they appear somewhat wavy; and, instead of slowly drifting up or down, they seem to waver with the sound from the speaker, as if animated. Sound bars appear when some of the sound energy reaches the picture tube and is reproduced as light. The cause of this malfunction is seldom a defective tube, rather it is

The effect of sound bars is similar to hum bars, except there are many more of them.

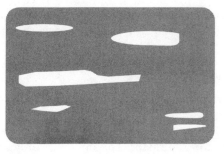

8-20 Sound bars are similar to hum bars, except that they change with volume.

caused by an improper adjustment—either of the fine-tuning control or, more seriously, of one of the tuned circuits inside the TV. However, only the fine tuning is within the capability of the beginner to correct. Adjust the fine-tuning control for a clean picture—even if it means some reduction in volume.

If this is not effective, it is an indication that either one or more of the IF circuits is improperly adjusted, or, in some TVs, the sound trap (4.5-MHz trap) circuit is improperly adjusted. This should not occur if the TV set ever functioned properly. It is extremely unlikely (although possible) that these circuits would drift out of adjustment. However, after a repair or alignment in the repair shop, some of these circuits might have been improperly adjusted.

Herringbone weave

A herringbone pattern on the screen seldom results internally, but these cannot be automatically discounted. One likely cause of the phenomenon is interference from another TV. Although the FCC has established some very clear standards for TV receiver radiation, many sets will make their presence known in other neighboring sets—especially in apartment houses and in master TV antenna installations. This can easily be identified by its random, intermittent nature and absence from most channels. It appears only when the offending receiver and the victim sets are tuned to the same station. Little that can be done about it, other than attempting to minimize it by touching up the fine-tuning control or switching to a different channel for the duration.

If the TV has an adjustable fine-tuning control, a herringbone pattern might be caused by improperly setting this control. In older sets, adjusting this control is required each time the channel is changed. The proper setting is the point that provides the best picture, not the loudest sound (these two were not always coincidental). In modern receivers, where the fine tuning is not critical and where usually one setting serves for all channels, there is considerable leeway in setting this control, but it can still be set (improperly) so as to produce a herringbone weave on one or more stations. An optimum setting (good for all stations, not best for any one) should be made by trial and error to avoid herringbone effects.

Incidentally, because many TV sets might receive the same channel on two adjacent selector positions (that is channel 2 on dial positions 2 and 3, channel 7 on positions 7 and 8, etc.), it is important to have the fine tuning optimized for the best results on the correct position. It is perfectly normal for some sets to have a herringbone pattern on the picture when, for example, channel 2 is examined on the channel 3 position of the selector. In such a case, the fine tuning is most probably incorrect.

Barkhausen interference

Barkhausen interference appears as one or more ragged vertical lines, usually in the left portion of the picture (Fig. 8-21). One line is always wider than the other lines. In tube TVs, some horizontal output tubes would oscillate and produce Barkhausen lines in the raster. The only way to get rid of the lines was to change the horizontal output tube or place a magnet around the middle of the tube. Regular Barkhausen magnets were designed to go over the output tube. By

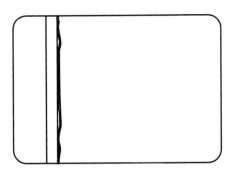

8-21 Barkhausen interference appears vertically to the left of the TV screen.

twisting or changing the position of the magnet on the tube, the lines would disappear.

Barkhausen (firing) lines in the raster of a solid-state TV chassis can be caused with a defective horizontal output transistor, flyback, broken or loose flyback metal core. Sometimes moving the flyback can make the lines to come and go. Replace the horizontal output transistor and snug up the flyback winding brackets, which hold the transformer to the chassis. Replacing the horizontal output transformer is quite costly and might be the only answer after all other tests are made. Replace the horizontal output transformer if it has a cracked or broken core. Be sure that the correct safety capacitors are installed.

Dark picture to the left

If a large shaded area appears to the left side of the screen, check for possible poor filtering of the boost power-supply circuits. This boost voltage supplies the video output transistors that drive the CRT cathode elements. If this symptom appears, check the boost-voltage source. A lower boost voltage will show the dark shading area or bars. Shunt the boost voltage filter capacitor with a another capacitor and notice if the lines disappear.

Vertical bars on the left side of the screen or picture can be caused by open or dried-up filter capacitors in the AGC or horizontal-output circuits (Fig. 8-22). Determine if the bars remain without a picture or antenna signal. Change the selector or tuner to a channel without a station. If the bars remain, the problem lies in the horizontal sweep circuits. Vertical lines that might vary with the signal, usually originate in the AGC or video amplifier circuits.

Several dark vertical bars on the left side of the screen can result from poor filtering in the B+ source that feeds the horizontal output transistor. Check for one or two small electrolytic capacitors in the flyback voltage source, close to the output transformer. These capacitors are connected in the circuit close to the primary winding of the flyback. Shunt each capacitor until the lines disappear.

If dark vertical bars cover the entire picture, check the vertical blanking circuits. Suspect that the blanking diodes in the video amp circuits are leaky. Test all diodes, transistors, and resistors in the vertical circuits. You might have to replace each one until you find the failed component.

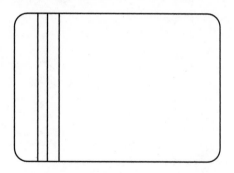

8-22 Dark vertical lines to the left can be caused by defective electrolytic capacitors in the AGC circuits and horizontal output winding circuit.

Picture ghosts

This malfunction is another one seldom caused internally, but it is quite easy to correct. A picture ghost is a duplicate picture, usually weaker than the original and usually displaced somewhat to the right. Figure 8-23 is one example of this type of ghost.

It is common knowledge that a TV signal travels from the transmitting antenna to the receiving antenna. It is also a fact, although not so commonly known, that the TV image is reproduced on the home screen from left to right; that is, the picture is generated much in the same sequence as the writing (or reading) of a line on a page. It follows that portions of the TV picture on the right appear later than the portions on the left. Of course, the difference in time is extremely small (millionths of a second) and the eye sees the whole picture simultaneously. However, if two signals from the same TV station arrive one behind the other because one comes in a roundabout way the second signal will arrive a little later. This is known as *multipath reception*—one signal travels the shortest path between the transmitter and receiver antennas, and the other follows a somewhat zig-zag path by striking some object on the way, then bouncing off, etc. This latter path is obviously longer, and takes more time. Figure 8-24 is a simple illustration of this occurrence.

If the TV receiver is far enough away from the transmitting center (such as a central metropolitan location) so that all stations lie in approximately the same line of path, the surest way to avoid or eliminate the type of ghost just described is to reorient the antenna. It probably has shifted so that it no longer points directly at the transmitters. Remember that TV transmitters radiate in all directions (omnidirectional), but all receiving antennas are relatively sharply unidirectional; therefore, it is possible that the receiving antenna now is aimed at a tall building or a hill that acts as a reflector for transmitted signals. In that case, the antenna is picking up a direct and a reflected signal.

If the transmitting location is not in sight, a map and compass should be used for correct pointing, although an aim-and-try (rotate the antenna, observe the picture, and repeat, if necessary) procedure is quite satisfactory. It is also fairly safe (although not absolutely certain) to assume that the correct direction is that in which most TV antennas in the immediate vicinity are pointing.

In cases where different TV stations are not in line, but are in definitely different compass bearings, an antenna rotor is essential. This will permit the receiving antenna to be pointed directly at the transmitter without any compromise.

As more and more TV receiving antennas are made to receive a combination of VHF (channels 2 through 13) and UHF (channels 14 through 83) signals, an additional adjustment might be required for elimination of ghosts and for maximum signal—especially on the UHF stations. This adjustment consists of the raising or lowering of the antenna, while simultaneously rotating it, for minimum ghosts and maximum signal. Contrary to most cases of VHF antenna installation, the highest possible location is not necessarily the best for UHF. In fact, lowering a VHF/UHF antenna system as little as a few inches might make the difference between a poor UHF signal and a very strong signal. Because the VHF is not so critical, it is safe to adjust

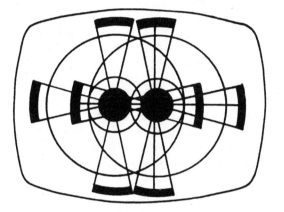

8-23 Ghosts are caused by multipath reception of two or more different signals.

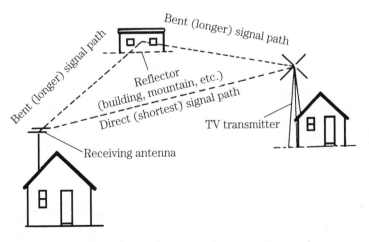

8-24 Multiple signal paths between the transmitter and receiver caused by ghosts.

the height for best UHF performance only. The direction of the antenna is fairly critical for all stations, so this should not be neglected.

Arcing in the picture or raster

Small white and dark lines going horizontally across the picture might be caused by outside interference or arcing within the TV. Disconnect the TV antenna lead-in wire to determine if the noise is being received by the antenna system. Automobile ignition, neon signs, power lines, and motors are a few of the man-made noises picked up by the antenna.

Arcing within the TV might be caused by the high-voltage anode connection arcing over. Disconnect the power cord and discharge the picture tube by connecting the picture tube's high-voltage anode connection to the chassis ground. Use a wire jumper attached to a long screwdriver that has an insulated handle. Slip the anode lead and clip it out. Notice if arcing marks are located under the rubber cover and check the clip to see if it is rusty. A poor ground connection between the outside area of the picture tube and chassis ground might cause small arcing that can be seen in the raster. Check and install a new ground, if necessary.

Defective yoke problems

Often, the vertical and horizontal yoke windings become shorted and present a trapezoid pattern on the raster. Shorted turns in the horizontal deflection yoke winding can produce chassis shutdown (Fig.8-25). Likewise, shorted turns in the vertical winding can produce improper vertical height. Smoke might be seen above the yoke assembly if it is arcing inside to another winding. Sometimes you can hear arcing inside the yoke assembly. If this is the case, replace the entire yoke assembly.

When a shorted yoke loads down the horizontal circuits and opens the fuse, suspect that the horizontal and low-voltage circuits are damaged. The output transistor might run quite warm with an overloaded yoke assembly. Remove the red yoke lead to the horizontal circuits. If the chassis remains on with a vertical bright line and no blown fuses, suspect that the yoke is leaky. Notice if the horizontal transistor is operating cool when the yoke is cut out of the circuit. Of course, the HV will be quite low with the red lead removed from the circuit.

Remove the yoke and check for leakage between the horizontal and vertical windings. It's difficult to locate a shorted turn or two with the ohmmeter because of the lower resistance within the horizontal windings. Check the HV with HV probe and notice if the voltage is quite low. Be sure that none of the pin-cushion transformers or connecting bypass capacitors are leaky connected to the horizontal deflection circuits.

Loss of fine detail

This defect might be caused by either component deterioration or by a misadjustment. Any portion of the picture where fine variations are present, such as a person's eyes or hair, a fine pattern on paper or clothing, and even the fine scan lines (raster) will be degraded by this defect. There are a few probable causes.

8-25 A leaky yoke between the vertical and horizontal windings can cause shutdown.

IF alignment incorrect

If the TV used to have good picture detail, a loss caused by IF circuit misalignment will seldom occur, except and unless those circuits have been improperly adjusted. The remedy is a complete IF alignment by a professional.

Fine-tuning control

Especially in very old TV sets, this can materially degrade picture detail because the ultimate effect is similar to that of a misalignment. The remedy here, too, is obvious, except that the setting of the fine-tuning control also affects the sound and, as covered previously, ghosts. An optimum, possibly compromised, setting is recommended.

Automatic fine tuning

Automatic fine tuning (AFT) is used in the latest TVs. When the AFT button is engaged, the sound, picture, and color will automatically track—even if someone plays around with the fine tuning control. Always keep the AFT button pushed in for normal reception. Each channel might be manually fine tuned while the AFT function is off. With the AFT function off, rotate the fine tuning control until the picture and sound is normal. Now, push in the AFT button and notice any change. If there is no change, go to the next channel and repeat the procedure until there is no change with the AFT button in or out.

Improper focusing

Although this misadjustment should be fairly obvious, the beginner might not be aware of it at all times, because, in mild cases, defocusing can be tolerated—especially in action scenes (which applies to most TV programs). Picture focus deterioration with long use is not unusual and it should be checked whenever picture detail seems deficient.

As detailed earlier, focusing is best done by closely observing the TV screen, using a mirror (if necessary), while adjusting the focus control for the sharpest horizontal lines on the screen. In older receivers, changes in the ion trap magnet position should be made with circumspection because it is easy to degrade picture brightness, picture size, and coverage of the screen edges by gross changes in the ion magnet position. Very slight and gradual rotation and sliding will easily showwhether the focus is affected.

Smearing

Smearing is best observed on the larger areas of uniform illumination, such as a wall, the border between a person's white shirt and dark suit, etc. Practically every cause listed previously for loss of fine detail applies here, with two exceptions. First, focus misadjustment is not nearly as evident on large areas of uniform illumination. Second, it is likely that a defective component in the video amplifier circuit is responsible for the loss of good quality in the heavy areas.

One other potential cause for loss of heavy detail is the misadjustment of the AGC. Because the AGC function sets the amplification level of the TV, it is, in a manner of speaking, a coarse presetting for the contrast control, while the latter is, comparatively speaking, a fine control. As stated earlier, the AGC should be set so that the contrast control is set about midway while receiving an average signal. Any setting of the AGC requiring the use of either the maximum or minimum extreme setting of the contrast control for an average signal is also likely to cause a type of distortion that ultimately evidences itself in picture smear.

A smeary-looking picture might be caused with open peaking coils or coupling capacitors in the video circuits. The picture is not clean, but the whole raster appears smeared. Check for an open peaking coil or delay line within the video circuits (Fig. 8-26).

Often, the voltage is quite lower than normal on the collector terminal of video transistor. These coils are very low in resistance and can be located with the DMM. If the picture smears intermittently, suspect that the coil winding is broken or has a poorly soldered connection. The defective delay line in the video circuits might cause video smearing. Check for a defective video IC if all other components seem normal.

Line pairing

Line pairing is a picture-quality defect that causes severe image deterioration, both in the fine detail and in the large solid areas, but particularly in the fine-detail area. It is caused by an accidental overlapping of adjacent raster lines. In appearance, the number of horizontal lines on the screen is effectively cut in half, while the thickness of each new line and the spacing between them is doubled.

In Chapter 1, it was established that the visible rectangle (raster) on the screen consists of approximately 480 interlaced lines; that is, lines "painted" by the electron beam, first the odd-numbered set (lines 1, 3, 5, 7, etc.), then the

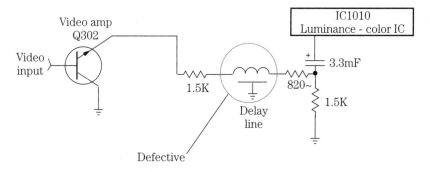

8-26 Open peaking coils and delay lines can cause smeary pictures.

even-numbered lines (2, 4, 6, 8, etc.). During normal operation, the sync or precise timing pulses from the transmitter keep the spacing between the lines exact, so that, for example, line 2 falls exactly midway between lines 1 and 3, line 4 fits exactly between lines 3 and 5, etc. In this manner, any picture information on one of these lines is clear and distinct from any information on an adjacent line. If, however, because of a malfunction, two adjacent lines overlap, two different sets of picture detail will overlap and neither will be clear. *Pairing of lines* refers to just such an overlap. A failure in the sync circuits, caused by either internal (TV) or external (freak reception or transmitter malfunction) causes will often produce this phenomenon. To verify whether or not the TV is responsible, a simple procedure is involved.

Switch to a different channel, preferably three or four channels away. If the malfunction persists, the fault probably lies within the set. Because it is extremely unlikely that two or more TV stations are having technical difficulties at the same moment, this source can be discounted. Similarly, freak reception, with possible interference from a more distant station on the same channel will produce just this effect, but is most unlikely to occur on a substantially different frequency. Thus, if the pairing occurs on channel 2, switch to channel 5 or 7 and observe. If freak reception is responsible, the other channels will most likely not be affected. If pairing persists, the TV is at fault.

Check the vertical hold control. Although this control setting is not very critical in the presence of a strong signal, a fine touchup might sometimes cure a mild case of pairing. Adjust this control for best interlace; picture rolling will automatically be corrected.

Check the AGC pot setting. The AGC adjustment cannot be overemphasized. It might seem paradoxical to the beginner and even to some TV servicer that too strong of a picture signal might go hand in hand with too weak of a sync signal. Because the AGC adjustment establishes the strength of the picture signal or contrast, in an effort to obtain a good strong picture where one did not exist (quite often for other reasons), the AGC setting can be advanced to the point where the sync pulses are reduced below a safe minimum. Adjustment of the control should be made until good sync performance (no more pairing) is restored—even if this produces a somewhat weaker picture, which can be corrected with the contrast control.

Missing adjustments

In a generalized description of causes and effects of TV malfunctions and their remedies, provision must necessarily be made for examples of controls and adjustments in a great variety of makes and models of sets. Frequently, therefore, you might find that your set has no such control or adjustment. One good example of this is the ion trap magnet. Although it is found fairly generally in older sets, such is not the case in more recent models. The same might be said for such an adjustment as horizontal lock or focus on black-and-white receivers. There are two explanations to this. In case of a "missing" control or adjustment, it is fair to assume that it is not required. Also, in such cases, the problem will not arise, at least not from this cause. In the other case, an exactly identical control or adjustment might not exist, although there seems to the nonprofessional TV do-it-yourselfer to be one with a similar name. Various horizontal adjustments are in this category. I have therefore provided in the various block diagrams a variety, in which the ultimate functions are the same, although the identifications are somewhat different—either because of small design differences or simply because of different terminology used by individual manufacturers.

The sync blocks in the various diagrams are a good example of this practice. Furthermore, in suggesting possible causes for one particular defect, I refer to these seemingly different stages for examination, indicating that they are functionally the same. Finally, in reading the instruction booklet or examining the various descriptions in the service information for a particular TV, you will be able to identify the exact function of the different component groups or blocks—even though the name might be slightly different from some typical TVs described here.

Shorts or fire

This type of failure, short or fire, in a TV set, although it might be relatively hazardous, is usually easy to diagnose, especially if the set stops working completely. There are a number of possible failures of this type, each of which requires a different procedure and remedy.

To prevent fire damage, always replace critical safety components, which have a star along side, with the exact part type. Capacitors that have a star along side might have a 5-percent rating with a certain working voltage. Bypass capacitors in the power supplies and horizontal circuits should be replaced with high-temperature, high-voltage replacements. High-voltage safety capacitors should be replaced with the same value and working voltage. These critical capacitors should be obtained directly from the manufacturer's service depot or distributor.

Choose flame-proof resistors where critical voltages and heat requirements are demanded. Make good, clean solder connections on the PC wiring. Poorly soldered connections of large-wattage wire-wound resistors (10 to 25 watts) can overheat and start a PC board fire (Fig. 8-27). Mount large-wattage resistors at least one inch above the PC board. Inspect soldered connections at on/off switches, ac power-line terminals, and surge power-line resistors.

Fuse blows

If the main fuse (the one that is accessible without removal of the back cover) fails, it might be caused by either a momentary overload, a weakness of the par-

8-27 Large-wattage resistors should be replaced away from wires and other flammable components.

ticular fuse, or an overload resulting from a defect in the TV itself. It is permissible to replace such a blown fuse with another one of exactly the same value. A lower value will blow without provocation, and a higher value might not protect the TV—it might even cause a fire. Carefully watch the behavior of the set immediately after replacement and for a while afterward. If no further difficulties arise (TV works normally, and you don't smell burning or hear crackling or frying sounds), you can assume that the fuse or a sudden surge in the line voltage was responsible, and the TV is not at fault (Fig. 8-28).

Replacement fuse blows

If the second fuse blows, whether immediately or soon thereafter, you must assume that the set is at fault. No further fuse replacement should be attempted.

Circuit breakers

What has been said about the main fuse of the TV applies equally to the circuit breaker—a device that has replaced the fuse in many TV sets. In simple terms, the circuit breaker is a type of lifetime fuse. Although the ordinary fuse blows during an overload and has to be replaced with a new one, the circuit breaker opens and disconnects the TV from the ac line. To restore operation, it is merely necessary to reset the circuit breaker by pushing a button on the outside of the device.

Two common types of circuit breakers are used today. One type can't be reset as long as the overload persists, so there is little danger of unknowingly causing the TV

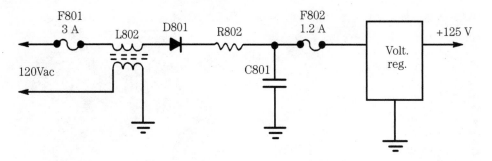

8-28 In this case, suspect components are overloaded in the regulator or horizontal circuits if F801, F802, and R802 are open.

to operate under damaging or even dangerous conditions. If a breaker can't be reset (the pushbutton does not respond to normal pressure), it is an indication that the overload is still there.

Thermistors

Many TV chassis have a thermistor located in the ac low-voltage power supply. A thermistor's internal resistance changes with its temperature, making it useful to control ac voltage. In older TVs, a 120-Ω (cold) thermistor was placed ahead of the degaussing circuits (Fig. 8-29). In later TVs, thermistors might be found with only a few ohms of resistance. A few seconds after current flow begins, the resistance of the thermistor will decrease to only a few ohms, letting the ac through to the power supply. If a TV with a good fuse or one with the circuit breaker reset is dead, suspect that a lead melted off the body of the thermistor. These components can easily replaced by unsoldering the two connections.

Burning smell, but fuse OK

Sometimes a fuse will not blow when it should. Fuses are not very accurate; their behavior is not 100-percent predictable. In such a case, damage to the TV is possible, depending primarily on how long the overload exists before the fuse goes (the correct and safe procedure is to immediately switch the set off) and whether it's a transformer or transformerless type set.

In transformerless sets, a local overload might be sufficient overheat or burn out a resistor. A burning resistor emits some intolerable odors. Although the remedy is beyond the scope of the beginner, the damage is seldom of major proportions. It is usually caused by the total failure of another component (a capacitor, for example).

In transformer-type sets, in addition to the same possibility as just outlined for a transformerless set, repeated blowing of the correct size fuse is very likely to be caused by a defect (short circuit) in the power transformer. It too might overheat and smell for a very short time (seconds, not minutes) without being permanently damaged. This, too, is a repair job for the professional. His or her advice on whether or not replacement is required must be taken.

As in the case of a burnout in a transformerless set, a failure here seldom takes other parts with it. In other words, a burned out transformer usually does not cause damage to other parts; however, as a matter of cause and effect, a transformer failure, in addition to resulting from an inherent defect, might also be caused by an earlier failure of another part. A defective or shorted silicon diode, filter capacitor, and high-wattage resistors might cause a blown fuse, open surge, or isolation resistors in the ac power line (Fig. 8-30).

In addition to a defective power transformer or shorted capacitors, there are many other potential causes of the fuse blowing—even if they are less common than these two. However, because (in all cases) the services of a professional are required, it is only necessary here to classify all fuse failures into two categories. One is a random failure not caused by any permanent defect. This one is identified by the fact that the correct replacement fuse does not blow. The second is a causative failure caused by a permanent internal defect or damage, which should be repaired by a professional.

In connection with random fuse blowing, it might sometimes be caused by injudicious placement of the TV. Because most TVs draw hundreds of watts, it is absolutely essential to allow the heat to dissipate. Pushing a TV cabinet tight against a wall or some other piece of furniture is sure to cause overheating. Regardless of whether or not the fuse blows, the heat buildup shortens the life of the many other components, not to mention the poor performance caused by thermal instability.

X-ray protection

In the latest TVs, an X-ray protection circuit prevents the chassis from operating when the high voltage has increased beyond the TV's limits. This protects the operator as well as the TV chassis. Excessive high voltage can result from an open safety or hold-down capacitor, defective regulation circuit, horizontal output, and flyback. If the high voltage increases, the current and voltage on the primary winding of the horizontal output transformer increases. This winding is fed from a x-ray shut-down circuit that cuts off the horizontal oscillator inside the deflection IC and shuts the chassis down (Fig.8-31).

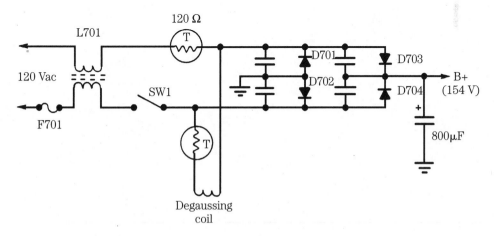

8-29 This thermistor is in series with the ac line and power supply. Another thermistor (120 W) can be placed in series with degaussing coil.

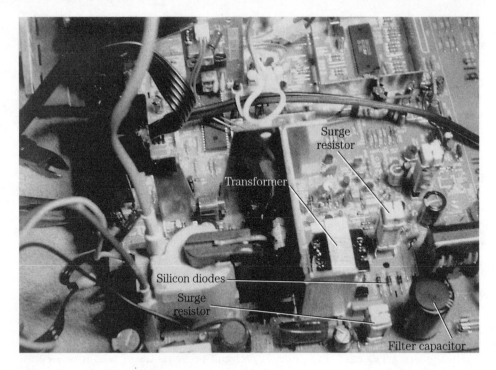

8-30 Shorted diodes, main filter capacitor, and large surge resistors could cause the PC board to burn.

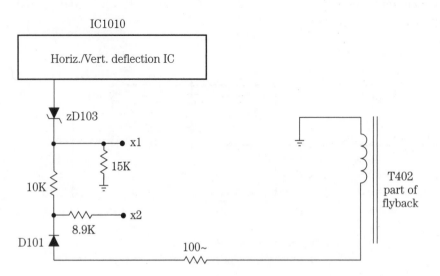

8-31 X-ray protection will shut down the chassis if the high voltage exceeds the limit, protecting the customer and components.

To check the HV shutdown circuits, be sure that the line voltage is around 120 Vac. Short terminals X1 and X2 with a test clip. The TV chassis should shutdown. Some sets will shutdown and automatically come back on in a few seconds. Most TVs can be started up again with out any problems. Check the x-ray circuits if the set does not shut down when terminals are shorted. Another method is to use a variac or step-up power isolation transformer, slowly raise the ac line voltage, and record the power-line voltage and high voltage applied to the picture tube when the chassis shut down. If the chassis shuts down before reaching the 120-Vac power line voltage, repair the horizontal or high-voltage circuits.

9
CHAPTER

Troubleshooting the color section

As has been pointed out, color and black-and-white receivers are essentially the same up to the video amplifier. In other words, all circuits and functions from the tuner through the video detector plus sync, sweep, and sound are common to both monochrome and color TV reception; therefore, the troubleshooting procedures covered so far apply equally to color and black-and-white receivers. In fact, color-TV troubleshooting, except for problems in sections that deal solely with the color signals, is easier when the color functions are disabled.

The early tube color circuitry contained a tube or dual-type tubes for each color stage; bandpass amplifier; color killer; burst amplifier; reactance control; reference oscillator; X and Y demodulators; and red, green, and blue color amplifiers (Fig. 9-1). Later each color tube was replaced by transistors and diodes. Today, all of the color circuits are included into one large IC.

Usually the chroma (color) and luminance (luma) signals are received by the TV antenna, selected with the tuner, and fed into a saw filter network. The output of the saw filter is fed into the IF and video detector stages. A 4.5-MHz signal is trapped out of the color-processing IC and fed back into the luma/chroma processing circuits, where the signal is separated from the B-Y, G-Y, and R-Y signals. All four signals are fed to the picture tube drive output circuits (Fig. 9-2).

You might find that one large IC contains the AGC, sound IF, chroma, vertical and horizontal deflection, and AFT circuits. Here, the TV processing IC only shows the circuit that corresponds with the color circuits. The B-Y, G-Y, and R-Y signals are fed to their respective color output stages and on to the color cathodes in the CRT. The luma or Y signals are fed to the emitter circuits of the color output transistors.

In another chroma/luma circuit, the input luminance signal is applied to pin 40 of the TV signal processor. Inside IC1150, the luminance signal is amplified and receives a dc voltage that is controlled by the sharpness control. A dc voltage applied to the picture or contrast control changes the contrast of the picture (Fig. 9-3). This signal is fed to the luminance control, where the brightness control controls the brightness

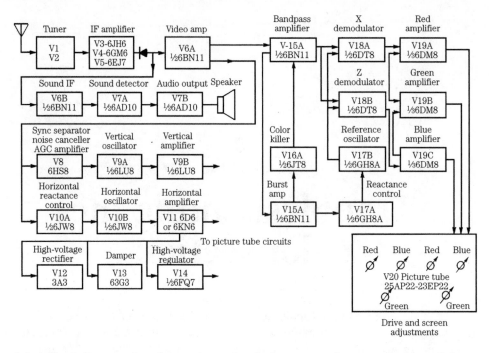

9-1 A block diagram of a color receiver using dual-purpose tubes as a color demodulator.

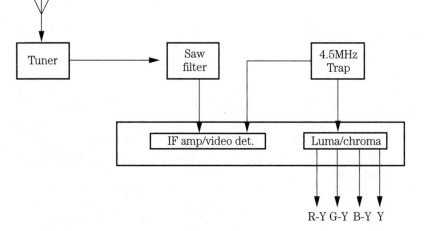

9-2 A block diagram of a color circuit inside a large IC (IC1001), which also has other stages, AGC/AFT/SIF/audio processing/sync separator/vertical, and horizontal deflection/X-ray protection.

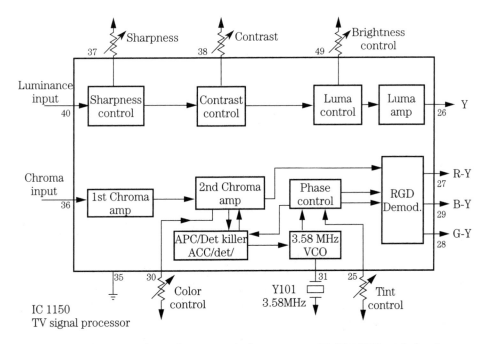

9-3 A block diagram of another TV signal-processing IC (IC1150) with luminance (luma) and chroma (color) circuits.

of the picture. The luminance amp amplifies the Y (luma) signal and is sent out to the buffer or luma driver. This Y signal is applied to the CRT board, where the Y signal is mixed with the R-Y, G-Y, and B-Y color-difference signals to develop the RGB signals. A vertical and horizontal blanking signal can be applied to the luma signal at pin 26.

The chroma (color) input signal is applied at pin 36 and sent to the chroma amplifier. This chroma amp, amplifies the signal and sends it to another chroma amplifier. The color level at the second chroma amp is sent by the color control with a dc voltage applied to pin 30. The output of the second chroma amplifier is sent to the RGB demodulator circuits, where the R-Y, G-Y, and B-Y signals are developed. The crystal oscillator (3.58 MHz) applies a VCO (variable-controlled oscillator) signal to the demodulator circuits through a tint phase-control circuit. The tint control applies a dc voltage at pin 25 to the test phase control.

The 3.58-MHz reference oscillator signal is applied to pin 31 and feeds to the VCO stage through the tint/phase-control circuits to the APC (automatic phase control) detector to compare the 3.58-MHz oscillator with the burst phase signal. This comparison voltage is fed to the 3.58-MHz VCO circuits. A constant color level output is controlled by the ACC (automatic color control) detector to control the gain of the chroma amplifier.

The chroma signals are processed inside the RGB demodulator, where the R-Y, B-Y, and G-Y signals are developed and found at pins 27, 28, and 29 of IC1150. The Y signals are applied to the red, green, and blue output circuits on the CRT board. They are mixed with the luma (Y) signal to drive the picture tube. Often, pins 27, 28, and 29 have about the same voltage and waveforms.

Symptoms

No color, but black and white normal

The first and most obvious reason for the absence of a color picture is the setting of the chroma or color input control. Because this affects all three primary colors, all color will disappear if the color control is accidentally turned off. A simple check of the control setting will determine whether or not it is the cause of no color. If it is, you can simply the control (assuming that no other adjustments have been changed) by eye, while observing the screen.

Fine tuning and color level

In addition to the obvious matter of setting color level from no color to any desired intensity by the chroma or color control, another factor affects the presence and intensity of color—the fine-tuning control. As covered in Chapter 4, reception of color depends also on the burst signal from the station and the 3.58-MHz color oscillator (block J, Fig. 4-1) in the receiver. Because the setting of the fine-tuning control naturally affects the received signal, it is possible, especially in a weak station, to have the fine-tuning control off its optimum position. Although a strong picture is still received, the color will be absent or very faint. Adjusting the fine-tuning control will bring the color component information up to an acceptable level. Because in all but a few TVs, fine tuning is preset for each individual station, such an adjustment will not affect any other station.

Color circuits

The first color amplifier receives the chroma input signal, amplifies it, and applies the signal to the second chroma amp. The amount of color in the picture or color level is adjusted by a dc voltage applied to the second chroma amplifier. This stage controls the amount of color applied to the picture tube.

3.58-MHz oscillator   The 3.58-MHz variable-controlled oscillator compares the APC detector signal, and the difference between the phase control and 3.58-MHz oscillator provides a reference signal. No color is in the picture when the 3.58-MHz VCO circuit is not functioning. The same is true with a defective crystal. A quick scope waveform at the signal oscillator indicates if the stage is oscillating.

3.58-MHz crystal   A brief explanation should be helpful. As mentioned earlier, certain very precise timing (or sync) signals are sent from the transmitter for cuing or timing the various signal sequences on the receiver. One of the signals exclusive to the color TV is the sub-carrier burst signal. This is a precise timing, and it is commonly called the *3.58-MHz reference*. Without the reference signal, no color reproduction occurs. The heart of this oscillator is a electromechanical element known as a *crystal*. In some TVs, it plugs into a socket, but in the latest chassis, the crystal leads are soldered to the PC wiring.

Remove the 3.58-MHz crystal, by unsoldering the terminals from the PC wiring, after first noting carefully the appearance of the picture tube screen. Often, a very poor crystal will still allow some color to appear on the screen. If the removal of the crystal makes no difference, then the fault does not lie here; however, if the slight coloration previously seen does disappear, the crystal is probably defective (or intermittent) and should be replaced. Typically, when an inoperative 3.58-MHz crystal is encountered, the screen will appear to have only blue and green—no red. Also, the tint or hue control will have no effect on the color present on the screen.

This is a standard item found at most radio supply houses catering to the TV servicers; therefore, a replacement crystal of the same type is readily obtainable. Where the service data lists a crystal by a particular manufacturer's part number, it might be more practical to obtain a replacement from the service department of that manufacturer.

In most recent TVs, the 3.58-MHz crystal is a pigtail type; that is, it is not like the plug-in type just described, but it is intended to be soldered in the circuit similar to a resistor or capacitor. In such a case, the crystal is "worked around" in the troubleshooting procedure until all of the other alternatives are exhausted. If none solves the absence of color, assume that the crystal is at fault and call a professional servicer. (See chapter 10 on do-it-yourselfer procedures on soldering printed circuit components as an alternative to calling a TV servicer.)

Color killer setting   In older color TVs, a color-killer control is adjusted to eliminate color lines, specks, and streaks from the black and white picture. Sometimes it is called *color snow*. Rotate the color killer control until all traces of color disappear from the picture. Today, the color killer circuits are controlled inside the color IC with no outside control.

RGB demodulators   The second chroma amp signal is applied to the RGB demodulator circuits, where R-Y, G-Y, and B-Y difference signals are developed. The chroma signals are processed inside the RGB demodulator with the -y signals at the output pins of IC1150. The R-Y, G-Y, and B-Y signals are fed to the base terminals of each color output transistor. This amplified signal is sent to respective cathodes in the picture tube.

Component failure   If all the preceding procedures fail to restore the color to an otherwise-normal TV, the fault must lie in one of the internal components in the path of the color signal, such as a coupling capacitor or a failure of a supply voltage to one of the tubes. In the first case, the tubes function normally, but the signal path is interrupted. In the second case, one or more tubes are inoperative because of the failure of the correct operating voltage. Whichever the case, the services of a professional servicer are required.

Simplified solid-state TV block diagram

The simplified solid-state TV block diagram consists of a varactor, band decoder, and multiband tuner. The tuner can be controlled with a microcomputer, system control, pushbutton, or remote-control systems. The IF amp from the tuner is amplified and passes through a SAW filter to the video amp stage. AFT, RF AGC, IF AGC, sound IF, video amp, sync separator, luminance, and chroma processing are located in one large IC (Fig. 9-4). The chroma amp and demodulator circuits connect to the CRT bias or color output transistor. The output from the three color output circuits feed to the cathodes of the picture tube.

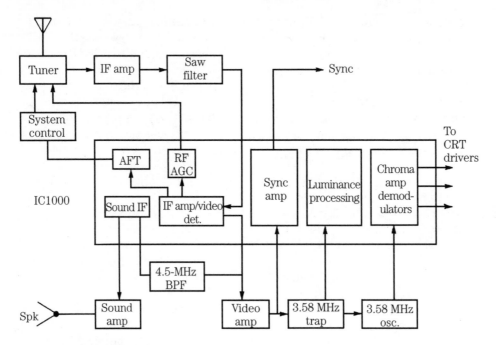

9-4 A simplified block diagram.

In the early solid-state chassis, all color stages contained transistors. Then, the novel IC was added to the color circuits. At first, one small, separate IC contained most of the color stages (Fig. 9-5). The first and second chroma amp, burst amp, ACC detector, R-Y demodulator, B-Y demodulator, killer amp, 3.58-MHz color oscillator, and color matrix were contained in one IC, along with the luminance components (Fig. 9-6). The red, green, and blue color output transistors were connected between the color IC and the picture tube.

First and second chroma amps

The chrominance signal is amplified by the first and second chroma amp and sent to the burst amplifier. Usually, the first chroma amp gain is controlled by the automatic chroma control (ACC). The color control proceeds the external bias to vary the voltage and gain of the second chroma amp.

Burst amp

The color signal from the second color amp is applied to the burst amplifier. The tint control is connected to the burst amplifier, and it applies external bias and voltage to the burst amp, changing the current ratio and color tint.

Automatic chroma control (ACC)

The ACC circuit stabilizes the chroma output against fluctuation of the chroma/burst signal level. The peak detection of the burst signal controls the gain of the first chroma amp with a voltage. When the burst output is large, this voltage reduces the gain of the first chroma amplifier, thus reducing the burst (chroma output) and maintaining a fixed amplitude.

Color killer

The color killer circuits prevent colored snow or noise from appearing in the raster when a black-and-white picture is being televised. This killer circuit switches the second color amp on and off, according to the transmitted burst signal. The color-killer control was used in the early TV chassis; now the color-killer circuit operates automatically within the luminance/chroma IC.

R-Y and B-Y demodulator

The chroma signal is applied to the B-Y and R-Y demodulator circuits. Often, the R-Y demodulator and G-Y demodulators operate through a lag circuit and feed to mix in the matrix circuits. In some circuits, the matrix color feeds directly to red, green, and blue output transistors. In others, a red, green, and blue amp or bias stages are used between the matrix outputs and the red, green, and blue color output amplifiers.

Automatic phase control (APC)

The automatic phase-control circuit regulates the color oscillator frequency and phase to ensure correct color sync and to provide signals to the various control circuits. The APC circuit detects the phase difference between burst signal and the

9-5 Soldering the color pin terminals of a large IC to correct intermittent color reception.

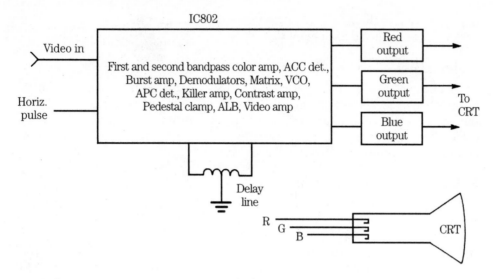

9-6 Today, most processing circuits including color, luminance, and demodulators have one large IC.

3.58-MHz oscillator, providing detection voltage. The detection voltage is controlled so that the VCO is in phase and frequency with the burst signal.

RGB output circuits

The color signal from the large IC can be fed directly to the red, green, and blue output transistors. In some later color chassis, the red, green, and blue amplifiers (plus the RGB output circuits) are also located inside a large IC. When the red, green, and blue output circuits are used separately, they are placed on the back of the CRT socket board. Actually, the RGB output circuits combine the color signal at the base terminals, and the luminance signal at the emitter terminals of each output transistor.

Color IC processing

Today, one large IC might include the AGC, IF amp, video detector, AFT, sound IF, audio signal processing, sync separator, vertical and horizontal deflector, X-ray protector, and luma/chroma processing circuits (Fig. 9-7). The video signal is applied to the bandpass filter, allowing only color to enter the first and second color amplifiers. The gain of the first color amp is controlled by the automatic color control (ACC) circuits. The gain of the second color amp is determined by the color-control dc voltage developed by AIU microprocessor. The second chroma amp is turned off and on when a black-and-white signal is received by the color-killer circuits. The output of the second color amp is connected to the color demodulator circuits.

The VCO color oscillator circuits are used when demodulating the chroma signal to the chroma demodulators (Fig. 9-8). The frequency is controlled by a 3.58-MHz crystal oscillator. The APC circuit compares the phase of the color oscillator to the

9-7 One large IC contains several different processing circuits.

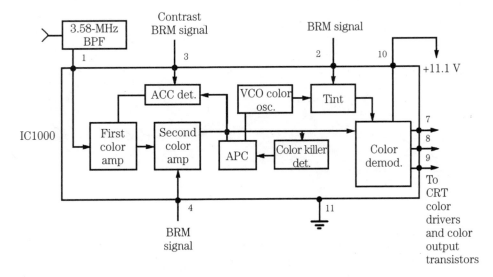

9-8 A block diagram of a late-model color-processing IC circuit.

color-burst signal, generating a phase error signal. This error phase signal keeps the color oscillator locked to the burst signal.

The chroma demodulator circuit uses the dc tint control voltage and color signal from second color amp to develop the R-Y, G-Y, and B-Y signals. The color-difference signals from the chroma demodulator circuit are fed to the CRT color driver and amplifiers, where the color and luminance signals are mixed and fed to the cathode elements of the picture tube.

Color modules

Color modules were used in many color TVs when modules were used throughout. Some RCA chassis had two separate color modules. In the Zenith chassis, the color circuits were also included with the AGC, IF, and sync circuits (Fig. 9-9). When the picture has no color, intermittent color, or mixed colors, simply replace the color module. Today, most of the color circuits are soldered directly into one large PC board, although one large IC may include all of the color, luminance, IF, AGC, and sync circuits.

Comb filter

The comb filter circuit separates the luminance (brightness) and chroma (color) video information so that when displayed on the screen, the cross-color lines are eliminated (Fig. 9-10). The video signal is supplied to the comb filter luminance processing channel and amplified by a gain stage. This composite video signal is first

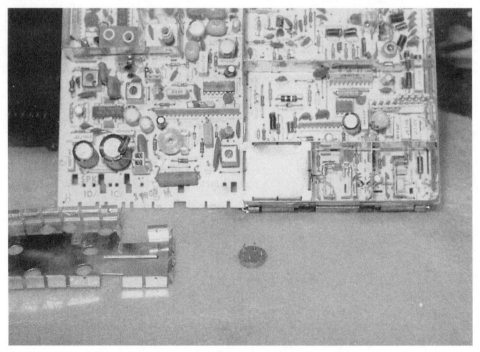

9-9 The TV might contain one or two color modules.

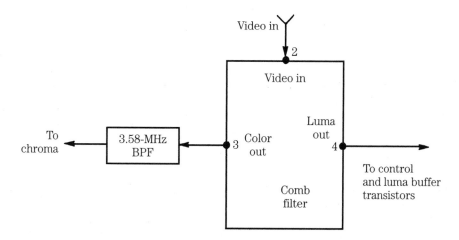

9-10 A block diagram of a comb-filter circuit in the luma/chroma circuits.

inverted, then amplified. The combed color output signal of the comb filter IC goes through a chroma-peaking circuit to the luma/chroma processing IC and the other path through a vertical filter network.

Other comb filter circuits might consist of a delay line driver, delay line, luminance buffer, luminance inverter, luminance equalizer, and peaking buffer (Fig. 9-11). The output of the peaking buffer transistor connects to the peaking circuit of the chroma/luminance IC. The luminance signal from luminance equalizer transistor feeds to the contrast amp. The luminance inverter provides a color signal to the chroma processing of IC650.

Weak colors

If all the colors are more or less uniformly weak, that is, no one color is much worse than the other two, the fault again lies in a common color circuit. The troubleshooting sequence here is logical.

Bandpass amplifier

A weak transistor in either of the bandpass stages, if there are two, will most likely decrease the color contrast.

Color control setting

If the color control has not intentionally or accidentally been backed off, in other words, if this control is in its usual position or setting, try advancing it clockwise. Advancing the color control will probably restore the color to a good level. Incidentally, it should not be necessary to have this control beyond its approximate mid-position, if the bandpass amplifiers are in normal working condition.

Color-killer setting

An improper setting of the color-killer control can produce the same symptoms as weak bandpass amplifiers. Unless you are sure that the setting of this control is

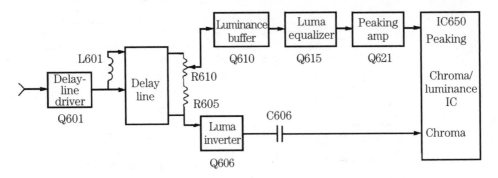

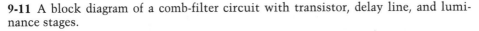

9-11 A block diagram of a comb-filter circuit with transistor, delay line, and luminance stages.

correct from previous observation under normal operations, the setting of this control might be responsible for reduced color performance. A quick check on the proper setting is to switch to a station transmitting no color at the moment and adjust this control, as outlined previously for the normal setting, to obtain a clean black-and-white picture.

Color-killer adjustments

In the early color TV, the color-killer control was located on the rear chassis or within the area of the color components. Tune in a black-and-white picture and adjust color killer control until the colored snow or noise is eliminated. No color-killer controls are in present-day color TVs. The color-killer circuit operates automatically within the chroma circuits.

Basic color troubleshooting

Take critical voltage and resistance measurements at all terminals of the color IC (Fig. 9-12). If the power supply pin voltage is low, suspect a defective color-processing IC. Be sure that the power source voltage is normal. With only a VOM or DMM, it is very difficult to know if the color oscillator is oscillating. Only a critical voltage test at each color IC terminal can help you to detect a leaky capacitor or a change in resistance.

By scoping the input, oscillator, and output circuits with the oscilloscope, you can quickly determine if the color IC is leaky or defective. Be sure that all voltages are way off before removing and installing a new color-processing IC. Remember, the open IC might have no voltage changes and replacement of the suspected IC is the only solution.

A lot of color problems within the color circuits can be indicated with measuring for improper voltages. An intermittent bypass capacitor can produce erratic color in the picture. A change in voltage on the color-killer bypass capacitor indicates low voltage that kills the color signal in the IC (Fig. 9-13). If the voltage at pin 5 falls below 5 volts, the color will disappear. If you are in doubt about any bypass or electrolytic capacitor, simply replace it.

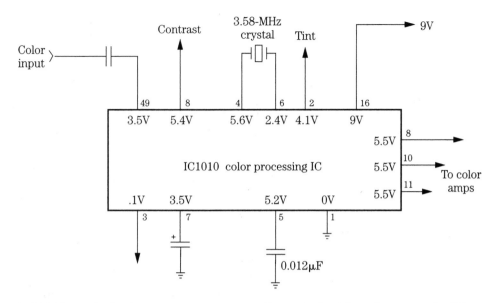

9-12 Take critical voltage measurements on all terminals of the color-processing IC to determine if the IC or connecting parts are defective.

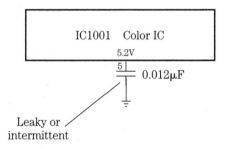

9-13 A leaky bypass capacitor can cause a dead or intermittent color problem at color-killer terminal 5.

Color functions

After the color detection occurs, the three primary colors exist individually, up to and including the picture tube. Thus, when it is stated that one color is absent, the various shadings or combinations of this color are also missing. A missing red signal denotes a picture consisting of the remaining two colors only. This implies that the absence of blue would show no shades of blue and also an abnormal green because the green signal is largely dependent on the existence of a normal red and blue. This might not follow from the logical understanding of colors in nature because it is based, instead, on the technical structure of the color system in the TV.

In color circuits with only one IC for the entire color circuits, except the color output transistors and picture tube, an improper power-supply voltage can eliminate color from the picture. A defective 3.58-MHz crystal results in no color. An improper setting of the color-killer control can cause erratic or no color, provided that one is located outside of the color IC circuits. Low voltage at the color-killer bypass capacitor terminal can cut off the color in the picture. Often the IC symbol used on the main TV schematic does not have the internal color circuits or block terminals identified (Fig. 9-14). You might have to identify the color pin terminals with the crystal, color and tint controls.

Besides the color IC, a missing color can result from a defective color amp transistor or picture tube. Poor or no color can be caused by a very poor black-and-white picture. Be sure the black and white picture is normal. A weak or gassy CRT can cause a very poor color picture. If one color is weak, suspect that a color demodulator circuit, color amps, or CRT gun assembly is defective. A blotchy-colored raster can be caused by a weak picture tube. Of course, the black-and-white picture will also be poor. Remember, a weak or intermittent color problem can be caused from a defective electrolytic capacitor in the color circuits.

Returning to the names and functions of the various stages, the demodulators serve to extract or separate the color signals from the combined electrical signals, which carry the color information through the TV. After the demodulators, the signals are the ultimate color information, which in some TV sets requires only additional amplification before application to the three color guns of the picture tube. As to the color identification, both the demodulators and the following amplifiers carry the designations R-Y, B-Y, and G-Y.

Notice that the hyphens between the letters are intended as minus signs and should read G minus Y, R minus Y, etc. These names describe a method of color processing that is technically called the *difference color system*, that is, difference color demodulators and difference color amplifiers. In the beginning, at least, it is not necessary to know the technical reasons for this. Instead, it is perfectly satisfactory to assume that R-Y stands for red, B-Y for blue, and G-Y for green. In no TVs will there be a G-Y demodulator because this is achieved as part of the R-Y and B-Y functions. Furthermore, in some TVs, the demodulators are marked *X-demodulator* and *Z-demodulator*. The first is equivalent to the R-Y demodulator and the second to the B-Y demodulator.

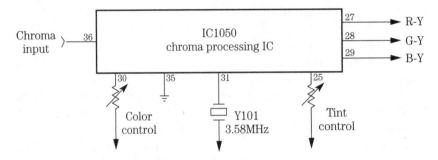

9-14 Internal block circuits show the correct terminals of the color IC on the main TV schematic.

Individual color adjustments

In addition to the previously mentioned overall color gain control, marked *chroma control* or *color*, there are two other sets of adjustments in the color-only portion of the receiver that must be taken into account when partial color deficiency is evident. One set of controls is called *drive*, the other set is called *screen*. There is usually one in each set for each color, namely, red drive, blue drive, and green drive, and red screen, blue screen, and green screen. Sometimes the red drive is omitted.

Drive and screen controls

Drive and screen adjustments are important to the overall representation of a color picture. Unless these setup controls are properly adjusted, it will be impossible to achieve a good color representation of the color telecast. Although these two sets of controls seem to have a similar effect on the color quality, their purpose is in two different extremes. Drive controls govern the highlights, and the screen controls set the operating point for each gun. That is, misadjusted screen control will have a more dramatic effect on the picture than a misadjusted drive control (Fig. 9-15).

Improper adjustment of the drive controls might produce, depending on whether the adjustment is too little or too much, either too faint a picture in the dimmer portions of the scene or too strong on the brightest portions of the scene. By comparison, the screen adjustment can make the whole picture look either too low or weak (insufficient) or too high (oversaturated). The normal adjustment for these twin-type settings is, first, to get the complete range of variation from dimmest to brightest, and second, to get this complete range up to a level that is a acceptable.

Gray-scale adjustment

The previous method for adjusting these two sets of related controls is based on purely subjective criteria—you adjust the controls until, to the best of your knowledge and observation, the color picture is most acceptable. A somewhat more impartial method, and one that the professional servicer usually follows, is what is known as the *gray-scale adjustment*. This simply means that the three guns of the picture tube are balanced to produce as nearly a perfect black-and-white picture as the design of the TV allows.

As explained earlier in connection with the characteristics of light, white light is a balanced combination of all colors. In a color TV system, this is modified to read: A normal black-and-white picture is the result of the correct combination of the outputs of all three color guns. The purpose of the drive and screen adjustments is to produce such a correct combination. The term *gray-scale adjustment* is borrowed from photographic terminology and denotes a continuous range from pitch black through all shades of gray to full white. A normal black-and-white photograph should show all these levels, provided, of course, that they existed in the original scene. In a three-gun color picture tube, proper setup of the gray scale is a prerequisite for a balanced color picture. But even in a gray-scale adjustment carried out by a professional servicer, there is a certain amount of magic involved, except in those cases of more recent sets, where a detailed adjustment procedure, applicable to a particular set only, is provided by the manufacturer. In such cases, there is often an auxiliary

9-15 The color screen and bias controls are on the rear of the TV chassis.

device marked *normal-service switch* (or similar terminology), indicating that the switch is left in the normal position for the use of the set (some normal functions are disabled in the service position), and temporarily switched to service for adjustment purposes. Even in these cases, the final touchup is still done by the servicer by magic. Based on all these factors, the best procedure for the beginner still is to use careful observation and judgment:

1. Turn the receiver on.
2. Turn the color or chroma control to minimum. If the set is equipped with a control that varies the gray-scale adjustments, set it to midrange. Some brands of receivers call this control *picture quality* or *tint*. Do not confuse this control with what is called the *tint control* on some makes. In one case, it is the hue control, and in the other, it is the raster coloration control. If it varies the coloration of the raster with the color control at minimum, it is the control in question. Set it to midrange so that when this procedure is completed, you will have the desired and intended range of the control.
3. Adjust the front-panel controls for normal viewing with the exception of the foregoing controls. Tune in a medium-strength station and fine tune the tuner.
4. Vary the brightness control from minimum to maximum, observing the entire screen for discolorations or tinting. If there is a normal black-and-white picture from low brightness to high brightness, it is not necessary to go any further. If this is not the case, proceed to step 5.

5. If the TV is equipped with a normal-service switch, as described earlier, set it to the service position. If the receiver isn't equipped with this switch, the procedure will be slightly more difficult, but it can be accomplished with a little extra care. With the switch set to the service position, a horizontal white line will appear.

6. Set all of the screen controls to minimum, noting their original position.

7. One at a time, adjust the red, blue, and green screen controls so that each color just begins to appear on the screen. Adjust all three so that each gun is turned on at approximately the same level. This is best accomplished in a darkened room because red will not appear so bright as the other colors and can be more readily seen in this manner. In receivers not incorporating a normal-service switch, you must "eyeball" the screen at a low-brightness setting to determine when each color just appears on the screen. The screen controls are set so that they have the same threshold level for each gun.

8. If the receiver is equipped with the normal-service switch, set it to the normal position. If it is the type without the switch, set the brightness control to a normal viewing level.

9. One at a time, bring each of the drive controls down from a maximum position to a point that results in a true black-and-white picture. Most often, the red drive is not adjustable and the end result will show the blue at maximum and the green slightly reduced.

10. Check the picture for gray-scale tracking at all brightness levels. When you are satisfied that the job is as good as can be obtained, check the color presentation by turning up the color control.

Color interdependence

Although separate final amplifiers can be used for each of the three primary colors, there is a certain amount of interdependence between them. So much so that a failure in one color will probably affect them all because of the technical nature of the color image structure. As detailed earlier, most TVs have two color demodulators, an R-Y (or X) and a B-Y (Z). The G-Y does not exist separately because it is, in a sense, a composite of the other two, as far as extraction of the signal is concerned.

I do not, of course, mean that a simple mixture of red and blue will produce green. What is meant is that the color transmission standards are such that proper blending of the R-Y and B-Y signals will also result in a G-Y output for feeding the G-Y final amplifiers. Furthermore, just as R-Y and B-Y (red and blue) components combine to produce the G-Y (green) component, a defect or malfunction in one of the first two will produce a corresponding deficiency or abnormality in the third one. For example, loss of the red signal will, primarily, remove virtually all the red from the picture and secondarily, distort the remaining color and particularly the green, into some other hue or combination of hues. A detailed troubleshooting sequence for the absence of a particular color follows.

Red, green, and blue color output circuits

The color output signal is fed from the luminance/chroma IC to the base terminal of each color output transistor. The color signal is coupled to the base terminal, and the luminance signal is tied to the emitter terminal (Fig. 9-16). Some TVs have color buffer or bias transistor stages between the color output transistors and the picture tube elements. Often, the CRT board contains the three color output transistors and required components.

The color output transistors have a higher dc voltage applied to the collector terminals (150 to 250 V) than most transistors. If one color is missing, either the color output transistor, components in the path to CRT cathodes, or the picture tube gun itself is defective. If the picture goes entirely one color (red), suspect that the red color output transistor or red gun assembly in the picture tube is leaky. Don't forget to check for dust inside a spark gap or defective spark gap if one color flashes on and off. Check voltage measurements and scope waveforms to locate defective components in the color output circuits. Usually, TO-220 power transistors are used in the color output circuits, and the collector voltage can be easily measured on the metal body of each transistor.

Red missing

Three areas affect the red color in the picture—the demodulators, the individual color amplifiers, and the picture tube. Based on some possibility of side effects on the other primary colors, as well as on the in-between shades, the test sequence is simple.

Blue and green appear fairly normal   Because you probably do not have access to a color-bar generator (except in the case of a built-in color generator), the judgment of color must be done subjectively. Look for known blue picture elements (such as sky) or known greens (such as grass) to judge. If such areas of known color seem to show normal color, the trouble is probably not in the demodulators.

Locate the red screen adjustment   This control affects only the red gun of the picture tube. Before manipulating this control, mark the starting position (knurled or slotted shaft) with a pencil or a piece of masking tape. Advance this control gradually,

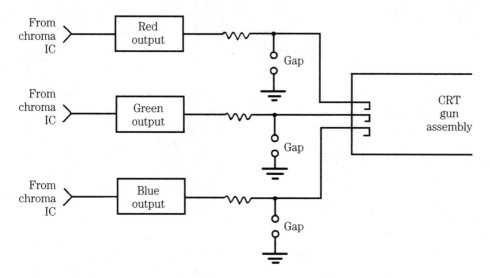

9-16 A block diagram of the red, blue, and green color output transistors.

usually clockwise, to increase the red color. If you see no noticeable effect, restore the shaft to the starting position. If gradual improvement results from turning the control, advance the adjustment further until the reds appear normal.

A note of caution: First, if the red screen control had accidentally been backed off (this is a possibility, but not a probability), advancing it to normal color production is perfectly proper. However, if the control has to be turned all the way to barely obtain a fairly good red, a defect elsewhere in the system is responsible and compensating for this fault with the screen control is not advisable for the good of the picture tube. Although it is quite possible that the red gun in the picture tube had deteriorated, this is not too common because, in normal use, all three guns deteriorate more or less equally.

Touch up the red drive control   This is important because the drive and screen controls for any one color are interdependent and might seem to act alike—especially in the hands of the inexperienced. As in the case of the screen adjustment outlined previously, it is best to observe the initial setting of the drive adjustment so as to be able to return it to its original setting if no change is apparent. The original setting of the drive control is important because, unless there is evidence to the contrary, the control was set for the proper operating conditions for the particular gun (red in this case). However, you're safe in making a slight adjustment, if there is clear evidence that such an adjustment, combined with a corresponding adjustment of the screen control, improves the red color sufficiently to make it acceptable. If such adjustment fails to show a substantial improvement in the red color, a defective component is probably responsible.

Blue and green missing   As in the case of red, the blue channel has its own demodulator (B-Y demodulator) and its final amplifier (B-Y amplifier). If the screen shows a color picture containing mixtures of red, but not blue or green, the indications are that a defect common to both blue and green is responsible because blue is one component of the green signal. The first and most likely suspect is the B-Y demodulator (or the Z demodulator, in some TV sets). The first step is simple.

If green is missing from the picture, the raster will contain a reddish-blue cast. Determine if the missing green color is at the output terminal of the demodulator IC, green color amp, or green picture tube gun assembly. A defective green color amp and picture tube gun assembly will show a poor black-and-white picture with the color-gain control wide open. If there is no color, but a good black-and-white picture, the problems are in the color IC circuits, not in the output or CRT circuits. Rotate the green screen control up and notice if any green color is in the picture or raster. Set up the picture with all screen controls for a normal black-and-white picture. Test the picture tube for a weak color gun assembly. All three color gun assemblies should be quite equal for a very good black-and-white and color picture.

Reduce the chroma gain

Reduce the chroma or color control to minimum so that no color is present. If the black-and-white picture is normal, the picture tube is okay and the fault lies in a defective component—an open capacitor or resistor. Of course, there is just the possibility that both the blue and green guns of the picture tube are defective, but you'll need a professional servicer for a positive diagnosis and repair.

Blue missing

A reddish-green cast indicates that blue is missing from the picture. If the blue waveform is found at the output of the B-Y demodulator, either the blue color output or blue gun assembly is defective in the picture tube. Turn the blue screen control wide open to verify that no blue is in the raster. Scope the blue pin of IC demodulator with no blue in the picture, with a normal blue output transistor.

Blue screen adjustment　Adjust the blue screen control on the back of the set. As covered earlier, a partial deterioration of any one color gun in the picture tube can be (temporarily, at least) compensated by advancing the screen control. If, however, only a slight improvement (or none) is realized from a major adjustment of the screen control, the fault lies either in a defective component in the blue signal path or (less likely, but still possible) the blue section of the picture tube is gone. In either case, outside help will be required.

On the other hand, if the blue screen adjustment produces a substantial improvement, a further correction can be attempted by a small adjustment of the blue drive control. Because this points to a deterioration of the blue gun, these two adjustments can restore the tube to a reasonably normal operating condition for as long as the tube will hold out. Again, this precaution is worth repeating: no decision to replace a picture tube should be made without the concurrence of a professional TV servicer.

Color presentation unstable

This could be either tearing of the color picture, similar to the diagonal zig-zag tearing of a black-and-white picture when the horizontal-frequency adjustments are off (see Fig. 2-12), or in less-severe cases, a partial breakup with the color wavering and general picture instability, often only intermittently.

It is assumed that the you are following the general admonition given earlier with regard to any color troubleshooting—before attempting any work in the color portion of the receiver, the performance in monochrome is checked. If the symptoms are the same as in color operation, the fault lies not in the color-only portion of the set, but in the common section, and repair should be made there. Only if the monochrome performance is normal, should work on the color section be attempted.

Try a new crystal

Where the second tube involved the 3.58 MHz oscillator, as in Fig. 9-4, replace the 3.58-MHz crystal if it is a plug-in type. It is also advisable in this case to observe whether the color breakup is always present from the moment the TV is switched on or if it appears later after the set warms up. If the latter is true, the crystal is a most likely suspect because these units sometimes stop functioning (oscillating) after a period of normal operation, but might return to normal if the set is switched off, then immediately switched on again.

Another possibility is poor contact at the crystal-connecting pins. As described earlier, in many TVs, the crystal connects to the TV by means of two thin wire pins, similar to those on miniature tubes. Applying a gentle pressure on the crystal case might momentarily restore normal operation if poor contact is the cause. In other TVs, the crystal is soldered in and replacement is not so easy (see Chapter 10). As a final note, a crystal failure or a 3.58-MHz oscillator failure usually, even if not always, causes a total loss of color information, not just breakup.

Color oscillator problems

The defective color 3.58-MHz crystal might cause no or intermittent color. Most color crystals are soldered into the PC board of the present-day TV. Before replacing the 3.58-MHz crystal, take voltage measurements on the oscillator IC pins (Fig. 9-17) Do not try to adjust the crystal frequency without special test equipment.

Often, pushing on the color IC or capacitors within an oscillator circuit might cause the color to become intermittent. Suspect poorly soldered IC connections if the color disappears from the picture and shortly returns without any control adjustments. Properly solder all IC chroma pin leads with a low-wattage soldering iron (Fig. 9-18). Sometimes spraying coolant on the chroma IC help you to locate the intermittent IC.

The best method to check oscillation of the color 3.58-MHz oscillator is with the oscilloscope. Check the waveform at pin 17 going to the VCO color oscillator. The color crystal is located in a metal case quite close to the chroma IC and oscillator adjustment (Fig. 9-19). The color wave amplitude might change if a color picture is being received (Fig. 9-20).

A final measure

If correct voltages are taken, the 3.58-MHz crystal has been replaced, and all voltages and resistance measurements are normal, but the TV still has no color, replace

9-17 Check the voltages on the color-processing IC if no color is in the picture.

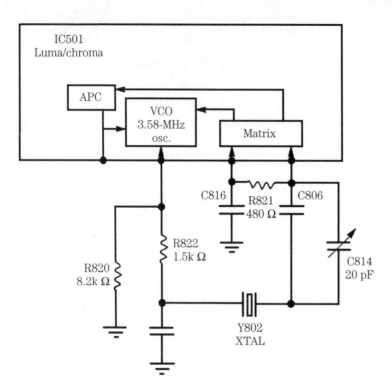

9-18 A block diagram of the VCO 3.58-MHz crystal circuit in one of the latest TVs.

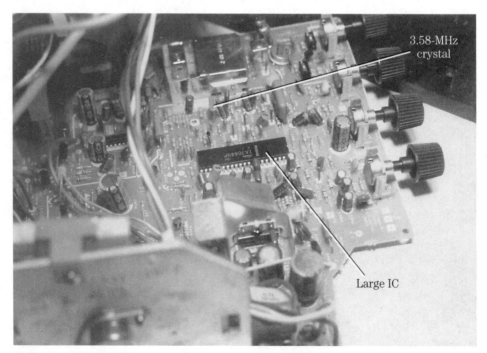

9-19 Locate the 3.58-MHz crystal to find the color circuits next to a large IC.

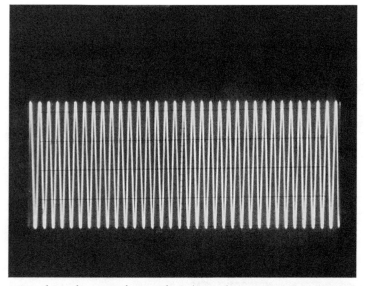

9-20 The color waveform taken from the 3.58-MHz circuit on pin 20 of Fig. 2-25.

the color-processing IC. If none of these steps seem to make an adequate improvement on color stability or normal picture, the fault probably lies in one of a number of components in the general area, requiring troubleshooting by a professional using specialized test equipment.

In a number of cases of mild color instability, the trouble lies outside the TV. Occasionally, a color is transmitted under circumstances making perfect stability unrealizable. The owner can best verify this by switching to an adjacent channel. If the problem disappears, the probabilities are that the station, not the TV, is at fault.

Color waveforms

To signaltrace the color circuits, the electronic technician should take critical oscilloscope waveforms. First, attach a color bar generator to the antenna terminals and turn on a color pattern. Scope the various pin terminals of the IC to see what section is dead or intermittent. Check the color input signal at pin 49 (Fig. 9-21). The input waveform should look somewhat like Figure 9-22. Check the color 3.58-MHz oscillator signal at pin 6 (as in Fig. 9-20).

Take a waveform test at output terminals 8, 10, and 11 for output signal of the color output transistors (Fig. 9-23). Check all three -Y color demodulator terminals. If one or all demodulator colors are missing, determine if correct B+ (V_{cc}) is present at pin 16. Last, check the flyback pulse at pin 22 (Fig. 9-24). If the input waveform, 3.58-MHz oscillator, flyback pulse, and B+ voltages are normal with no waveform at pins 8, 10, and 11, suspect IC1001 (color-processing IC) or the attached components are defective.

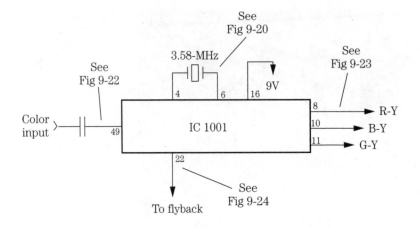

9-21 The various scope waveforms taken on the color IC to verify if color is present or missing.

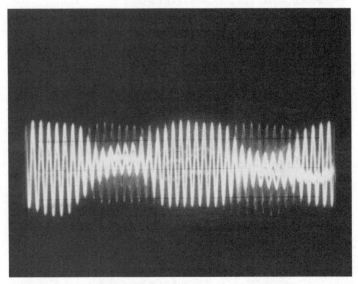

9-22 The color input at pin 49 of the processing IC.

Color problems

Intermittent color and no color are the most common problems in the color circuits. Drifting color bars might be caused by a defective component or misadjustment of APC alignment. If one color is missing, check the color output demodulator signal, color amp, and color gun assembly of the picture tube. Recheck the spark gap of the missing color. Sometimes if a defective spark gap arcs continually, the picture tube will shut down the whole chassis.

Accurate voltage measurements on the chroma IC might help you turn up a leaky component. Sometimes the chroma IC might be defective, but still have fairly normal voltages; it must be replaced. Scope the color input and output signals with the scope and color-dot bar generator connected to the TV antenna terminals. Remember, the color circuits must have a video and color input signal, usually from the comb filter, and an output color signal from the color demodulators to the color output amplifiers (Fig. 9-25). The VCO 3.58-MHz color oscillator must oscillate, and a horizontal pulse from the flyback circuits must be present before the color stages can operate in today's color TV. Of course, some chroma circuits might have sand- castle input and BRM signals applied to the chroma processor. It's best to leave the color circuit problems up to the experienced electronic technician.

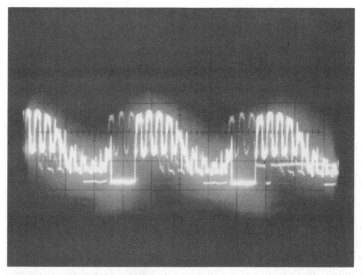

9-23 The color output waveforms taken from one of pins 8, 10, and 11 of the demodulators.

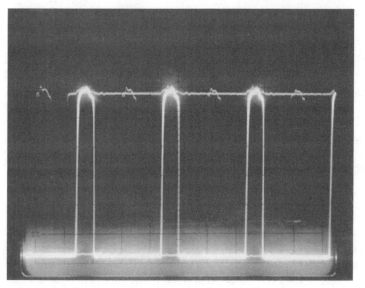

9-24 The flyback waveform taken on pin 22 of IC1001.

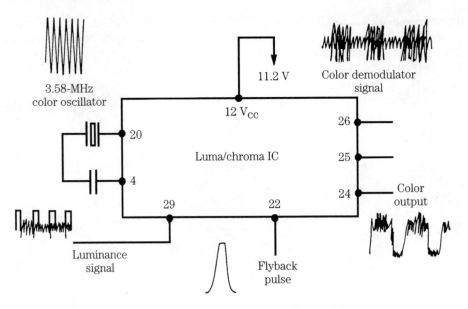

**3.58-MHz
color oscillator**

11.2 V

Color demodulator
signal

12 V$_{\text{CC}}$

20

Luma/chroma IC

26

25

4

24

Color
output

29

22

Luminance
signal

Flyback
pulse

9-25 Signals and scope waveforms taken on a normal color IC.

Antennas

Commercially available TV antennas vary all the way from a piece of twin-lead wire cut to some approximate length, costing a little over a dollar, through various dipole versions, all the way to multielement, multibay structures, costing hundreds of dollars. The more elaborate the antenna, the higher the gain (a term indicating the strength of the signal delivered to your set), and vice versa. An important characteristic closely related to gain is the bandwidth.

The second factor, although must less scientific, is nevertheless just as important: viewer attitude. It is a fact that the average observer is far more tolerant when viewing a black-and-white picture than he or she is of color. This is because (in part, at least) even a faint picture can be seen and followed in monochrome, but a weak color picture intolerable and actually much more difficult to follow.

First, the satisfactory color antenna must have adequate gain. Although almost any antenna, given a fair to poor signal, will provide an acceptable black-and-white picture, only a sufficiently high-gain antenna will serve as well for color reception under the same circumstances. Second, the color antenna must have adequate bandwidth. That is, it must be able to intercept and receive a strong-enough signal over the whole expanse of the TV frequency spectrum, from channels 1 through 13, and possibly channels 14 through 83. Of course, this does not mean that a narrow-band antenna will receive only part of the picture. It will receive a single channel or possibly two or three. Narrow-band antennas are usually cut for specific frequencies.

The characteristics of an antenna can be obtained without great effort. First, a few consumer organizations do impartial testing and reporting on manufactured products, antennas among them. Their reports are published and can be easily obtained for a minimum cost, or they can be seen in most libraries. Second, and without even any factual data, and contrary to other examples, the cost of the antenna is a good indication of its capability. As a matter of confidence, the purchaser might rely on the reputation of the manufacturer, be it a TV manufacturer who has antennas made under its label, or the antenna manufacturer itself. The only precaution is to steer clear of the sensational "amazing-discovery" antenna "bargains" offered by (usually unknown) mail-order advertisers. The old cliché that good quality does not come cheap is very true here. In cases of color abnormalities, such as smear, color over-lap, and color snow, assuming that the set has otherwise been checked out, it is advisable to consider the antenna as the possible source of the trouble.

The foregoing observations are not intended to suggest that you should always buy the most expensive antenna. Not only is this unnecessary, but sometimes actually inadvisable for the following reasons.

The purpose of any antenna is to provide an adequate signal for the particular receiver location, not the maximum signal possible. A simple perusal of catalogs and advertisements will show that antennas are classified according to intended use; thus, there are antennas for metropolitan, suburban, near-fringe, fringe, and deep-fringe applications. Advertisements often specify mileage ranges for each type, but these should be taken with a grain of salt because they are for ideal conditions, not usually realized. A practical guide would be to purchase an antenna one grade better than indicated by the advertised mileage; thus, if the TV is located in the near-fringe distance from the station, a fringe antenna would be adequate. Of course, local conditions, such as locations behind a hill, etc., will adversely affect the signal. In such a case, see what size and type of antenna is used by others in the immediate vicinity. Having determined how big an antenna to get, it might further be advisable to get one from a reputable manufacturer (Fig. 9-26).

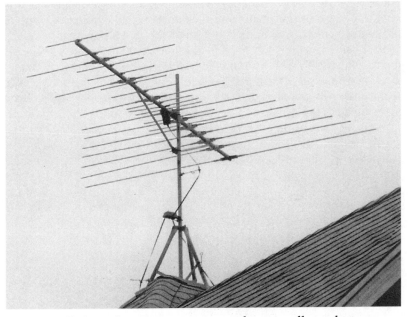

9-26 The all-channel VHF antenna mounted on a small metal tower.

Combination VHF/UHF antennas

In deciding on the type of antenna, consider the reception requirements in the UHF portion of the TV band, both for the UHF reception, as well as to the incidental effect on the more popular VHF stations. Because most network and independent TV stations of importance are concentrated in the VHF region, it is beneficial to decide, before installing an antenna, where your primary interest lies. If it is in VHF only, the best VHF-only antenna will be best because almost allcombination VHF/-UHF antennas are a compromise, with less-than-optimum reception in either range. If, however, the set owner is interested in the best reception of both types of stations, then a separate UHF antenna is desirable. The only exception to this situation is in areas of high-signal strength. Here, a combination antenna will be adequate for both VHF and UHF.

Color smear

In a black-and-white picture, a condition similar to smear often results from an improperly installed or oriented antenna. The usual name for such an appearance is ghosts—a slight displacement of the picture edge as if two superimposed identical pictures were slightly out of register or slightly displaced from one another—either to the left or to the right. One check on the accuracy of this diagnosis is to turn the chroma (or color gain) control all the way off so that a black-and-white picture remains. If evidence of ghosts exists, the diagnosis was probably correct. Adjusting (rotating) the antenna, if it is not otherwise defective (broken elements, poor connections at the antenna, loose elements, etc.) should eliminate ghosts or at least reduce them to a minimum, and markedly reduce the color smear.

Degaussing problems

A fairly common cause of both color smear and color mixing or distortion, particularly around the edges of the picture, is accidental magnetization of some of the structural metal around the face of the picture tube by stray magnetic fields: ac lines in the wall, a heavy-current appliance in the immediate vicinity of the TV, etc. Sometimes moving the TV to a different wall causes improper purity by changing the picture tube angle to the north pole. Stopping or starting an electric sweeper in front of the TV could magnetize a portion of the picture tube. Even placing strong magnets of the speaker column mounted next to the TV could magnetize the front of the picture tube. Often, demagnetizing or degaussing the picture tube will remove any color impurities from the screen.

Although most of the more recently manufactured TVs have built-in degaussing circuits, the earlier color sets are not so equipped. With the built-in degaussing circuit, the front of the picture tube is degaussed each time the TV is turned on. When power is turned on, the thermistor in series with the automatic degaussing coil (ADG) provides a low-resistance path to the degaussing coil (Fig. 9-27). The ac voltage is applied across the degaussing coil at the front side of the picture tube. The coil energizes and removes most color-purity problems from the picture tube. After a few seconds, the thermistor becomes warm and provides high resistance to the ac voltage, shutting down the degaussing process.

If a section of the screen is colored, shut the TV off and wait a few moments. Now, turn the TV on and see if the colored section is removed. If not, try degaussing the front of the picture tube with the external degaussing coil (Fig. 9-28). Although

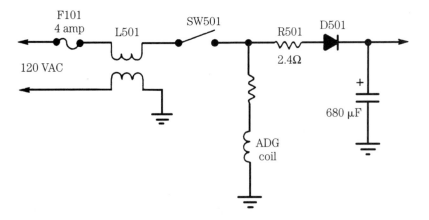

9-27 The degaussing circuit of the low-voltage power supply only operates for a few seconds when power to the TV is turned on.

9-28 Use the external degaussing coil to remove stubborn patches from the front of picture tube.

the process of external degaussing is relatively simple and quick, it does require a degaussing device, a large (10- or 12-inch diameter) ring consisting of many turns of insulated wire terminated in an ac plug.

Degauss the picture tube by plugging the ac cord into the receptacle and rotating the coil ring in a circular motion at the front of the picture tube. Be sure that the coil passes around the outside edge of the picture tube. A few minutes are required to demagnetize the steel frame and the rim of the color tube. To finish the degaussing process, bring the coil around to the outside edge of color tube, and while backing up, rotate the coil. Bring the coil away from the picture area by turning it perpendicular to the TV screen before unplugging the degaussing coil. The electronic technician would typically degauss the picture tube when finished with the repair of these older TVs.

Sometimes the degaussing coil or plug opens and no immediate degaussing occurs. Before long, a color section appears on the TV screen, usually at the corners. Check the thermistor resistance with the ohmmeter for open conditions, up to 25 Ω when cold. Examine the thermistor to see if it is open or has a "popped off" terminal, caused by overheating. An open ADG coil or plug results in no degaussing. Also, the coil can be checked with the ohmmeter and it should be below 100 Ω. In a few cases, the insulation of the degaussing coil has broken down and shorts against the metal framework. If the fuse keeps blowing, disconnect the degaussing coil (because of a possibly shorted condition) to prevent the fuse from blowing.

10
CHAPTER

TV repair procedures

Before listing some of the major differences between tubes and semiconductors, transistors and diodes perform the same functions as tubes; that is, they detect, amplify, oscillate, etc. The end result, whether picture or sound, is exactly the same, regardless of how it was achieved. Nevertheless, the structural and operational differences between tubes and transistors are very sharp. Ones that are important to the consumer are:

- Unlike tubes, transistors need no heaters for their operation. This is perhaps their greatest single advantage, resulting in greater stability, longer life, and more economical power consumption.
- Transistors are low-voltage devices. Most operate on between 1.5 and 15 volts, although a few rare ones require as much as 100 volts or so. By contrast, tubes usually require a few hundred volts.
- Transistors are not fragile. Shock and vibration are far more harmful to tubes than transistors. This relative immunity to vibration combined with low-temperature operation makes for long life.
- Transistors are simpler to manufacturer and hence less expensive.

Effect on TV design

Solid-state technology has produced a number of changes in the TV industry. To the do-it-yourself consumer, most of these changes have been positive. Although there might be a few disadvantages, as far as repair by the owner is concerned, on the average, the consumer is ahead. Some of the more obvious advantages are:

- *Smaller size* With the proliferation of semiconductor use, a major miniaturization has evolved for all components except, of course, the picture tube and the speaker. A by-product of the parts miniaturization has been a decided improvement in the quality of components.
- *Lower heat* Not only has the power consumption been greatly reduced, but heat generated also has gone down correspondingly. Less heat means

less waste, but far more important, it means less wear and longer life for components. This is almost entirely because of the replacement of tubes with transistors, although some progress in this direction was independent of the transistor factor.

- *Less wiring* One of the greatest and most obvious effects of the solid-state technology has been the tremendous reduction in wires and wiring. And although wires are not usually a source of trouble, their elimination wherever possible is a distinct plus for economy of space, labor, and cost.
- *Modularization* With parts being much smaller, it has been logical to group a larger number of them on a single common subassembly; thus, printed circuit modules or boards came into vogue. More information about these devices is listed in the section describing do-it-yourself techniques.
- *Repair standardization* When TVs consisted of almost 100-percent discrete parts, no two jobs were alike, that is, the manner of repair, and to some extent the efficacy of such a repair, was largely dependent on the individual TV servicer, professional, or do-it-yourselfer. The PC board and module have eliminated a lot of haphazard repairs. For any particular repair, the replacement module is connected exactly the same way as the original—the repairman has no choice in this matter. It will produce essentially the same normal operation of the set, regardless of who repaired it.

At first glance, it might seem wasteful to replace a complete module when only one of its components is defective. Actually, this is not necessarily true. Although it is sometimes possible to quickly locate the defective part and replace it only, thus saving the old board, in the majority of cases, the replacement of the complete module is more economical. It saves time, money, and possibly even avoids the risk of damage to other components in the immediate vicinity of the defective part—a most unpleasant experience considering that the damage might not be discovered until after reassembling the TV.

Service techniques

There are a few precautions as well as some different techniques to be observed in working with solid-state circuits:

- You must be careful not to disturb anything in the vicinity of the part being removed.
- An absolute minimum of heat must be used in soldering components.
- Physical force, where necessary, must be applied very sparingly. Miniature parts simply cannot tolerate rough handling.

The following few tools are recommended as an essential minimum in servicing solid-state receivers:

- Needle-nose and long-nose pliers, small size.
- Diagonal cutting pliers (dikes), small size.
- Pencil-type soldering iron, not over 30 watts.
- Pair of tweezers, about 6 inches long.

- Dental probe (it looks like a giant needle) with a flattened end for poking and scraping in tight quarters.
- Heatsink (a type of clamp that snaps shut in the manner of a clamp) to be used in absorbing any excess heat during soldering, thus preventing damage to parts.

One other tool, not exactly a must, but very handy, is a type of soldering tool that automatically removes solder by suction, incident to unsoldering small parts, transistors, etc. During such unsoldering, very tiny mounting holes used for parts mounting, unintentionally fill up with solder, preventing the new part from being inserted; however, with a bit of experience and care, the dental probe in conjunction with the soldering iron can accomplish the same job. Also, the unplugged hole can be carefully heated while a fine needle (not a wire!) is moved back and forth in the hole, thus dislodging the solder. Toothpicks work just as well if you can keep a supply on hand.

Required test equipment

The beginning electronic technician should have a good VOM or DMM for accurate voltage and resistance measurements. The digital multimeter (DMM) is a lot more accurate for small resistance and voltage measurements in the solid-state chassis (Fig. 10-1). A transistor tester is handy to test those transistors and diodes in and out of the circuit. The third test instrument should be a low-cost color

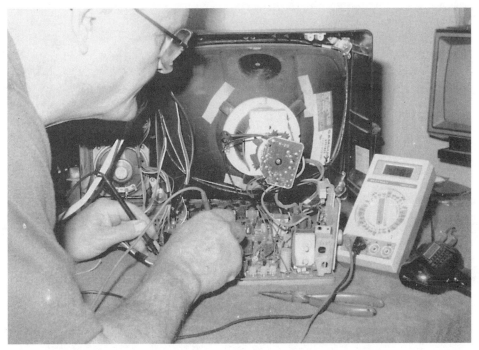

10-1 A good DMM and transistor tester are two important test instruments for the TV-repair beginner.

dot-bar generator to set up and adjust the picture tube, and remove and install the CRT. Next, purchase a good oscilloscope (40- to 100-MHz bandwidth) and capacitor tester that will test capacitors up to 2000 μF.

PC wiring and soldering

The printed-circuit technique eliminates old-fashioned point-to-point wiring by electrochemically printing the wires on a phenolic or fiberglass board. First, a layer of copper is chemically deposited on one or both (as required) sides of the board. Next, an etching process removes the copper from all areas, except those to be used as conductors or connecting points. On two-sided boards, the wiring of both sides can be connected together with metal feed through or griplets. Holes are then drilled for inserting the wire leads of the components to be mounted. Parts are then inserted in the proper holes and the board given a solder bath, which solders all leads to their respective copper backs or areas (Fig. 10-2).

In the manufacturing process, a machine inserts the small components while a person mounts the larger parts. After all components are mounted, the board travels through the solder bath (just the bottom side of the board) and onto a testing board. PC boards that have a defective connection are automatically rejected. Most of these discarded boards are repaired individually and passed through the testing station again. Sometimes, the human hand never touches the printed board until it is mounted in the TV.

10-2 An etched PC board with SMD components on the wiring side in a 13-inch RCA portable TV.

When mounting a defective part, heat is applied with a fine-tipped soldering iron to the immediate area of the joint. Usually, a heatsink is not required when removing defective components. A heatsink should be used when installing solid-state components because excessive heat will damage their internal connections. Special heatsink clamps can be used or you can use a pair of needle-nose pliers. Grasp the transistor terminal that you wish to solder with the pliers as you solder the terminals.

As in all soldering and much more so here, time is the most important. The longer the soldering iron is applied to the component the greater the risk of overheating and damage. With care and a little experience, a good joint can be made by a touch-and-go application of the iron. This method requires:

- A shiny, clean tip on the soldering tool.
- A corresponding clean spot on the work being soldered.
- A good contact between the soldering tool and the work.
- Choose a small iron (less than 40 watts) to solder transistors or IC circuits. A rechargeable battery-type soldering iron is ideal (Fig. 10-3).

Place the point of the iron on the junction of the wiring and the component terminal, and flow the solder from the opposite side. It's best to use a small-diameter solder in transistor and IC soldering applications. A little practice on a few scrap pieces of material can soon make one a proficient solderer.

Choose the correct type of solder for electronic work. Never use acid-core solder on any electronics. There is a great variety of solder on the market—from the coarse, lead-like stuff used by plumbers to a very fine, high-grade type used by the military for precise electronic equipment. It is quite easy (even for the novice) to know which is which. Solder (especially the kind used in electrical work) is rated in the percentages of lead and tin used in the mixture. The higher the proportion of tin, the better the solder, and the shinier it looks. Thus, a 40-60 mixture is a relatively poor grade with only 40 percent tin. 50-50 is commonly used in radio work, but is not recommended here. It is better to choose a 60-40 grade, that has a 60-percent tin content. Solder is manufactured in three different diameters or gauges. The two smaller-diameter types are best for transistor and IC applications. A pound of this type can cost a few dollars, but it should last the do-it-yourselfers for a lifetime. This type of solder has a resin core (a resinous chemical in its center for cleaning the surface during soldering), so it is advisable to wipe the soldered area carefully and sparingly with alcohol to remove any resin residue, which could (in some cases) cause electrical leakage. Stubborn resin residue can be removed from the area with the blade of a pocket knife or with a small wire brush.

Consider a number of factors before deciding whether or not to tackle a PC board. First and most important is the capability of the individual. This includes any previous experience, ability to work with small tools, as well as handling tiny parts, fine soldering, and general manual dexterity. Based on one's self-confidence with regard to these skills, you might decide to go it alone or to seek the services of a professional.

The second factor is the nature of the suspected or failed part. For example, an encapsulated sub-module is determined to be defective, removal of the part and replacement is far preferable to scrapping the large PC board on which the submodule

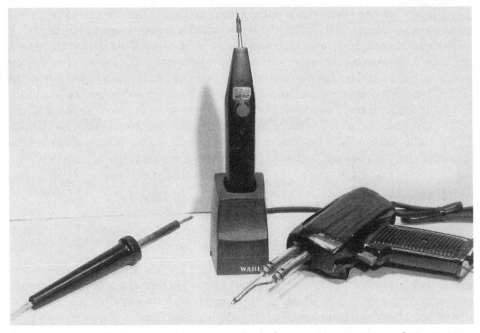

10-3 A rechargeable-battery soldering iron is ideal to use in transistor and IC circuits.

is mounted. On the other hand, if you suspect a relatively small circuit board that has parts soldered flat against the surface of the board with no accessible wires or leads, replacement of the whole unit is the wiser alternative. Finally, the cost of the replacement part is a factor. The purchase price of the part, as well as the repair shop bill, should be taken into account, the latter almost always being the larger of the two.

To summarize, if after careful examination of the defective part or parts and an equally careful study of the applicable service information, you think that you can do the job, you will probably be successful. If, however, after all these preliminaries, you decide that you are not up to the task, the job should be left to the professional TV servicer.

PC board parts layout

Today, most TVs have one large PC board with parts mounted on top. Often, the various components are clustered together with the different circuits. Within RCA's (1990) 13-inch TV, parts are mounted on top of the board with miniature surface-mounted components under the same PC board (Fig. 10-4), although several tie wires are on top of the chassis to tie the PC wiring together. In this particular chassis, PC wiring is used only underneath for the top-mounted regular parts and underneath with surface-mounted components. Other TVs use double-sided PC wiring both on top and on the bottom of the one board.

You might find it takes a little longer to locate the various components underneath and on top of the PC board. The manufacturer might include brief drawings indicating where parts are mounted. The manufacturer's service literature and schematics are must items in trying to service the solid-state chassis. Be extremely careful when removing and installing the small components to prevent breaking the board and the miniature parts around the replacement. Try not to flex or bend the large PC board or it might break the connections of the small surface-mounted components.

Replacing surface-mounted components

Surface-mounted components often appear as very small rectangular brown and black pieces with soldered ends. Bypass capacitors and resistors might appear with single or dual units within one component. The RCA CTC145E chassis (1990) has surface-mounted components underneath and on the same side as the PC wiring (Fig. 10-5). Standard components are mounted on the top side of the PC board. Here you will find thin lines of PC wiring connecting the surface-mounted and regular components together (Fig. 10-6).

Besides the TV, surface-mounted components are used in CD players and camcorders. Surface-mounted ICs and microprocessors have wing-type soldered connections that solder directly onto the PC wiring. Most ICs are marked with a dot or number that indicates the correct starting pin number. The small microprocessor might have up to 80 or more terminals (Fig. 10-7). Be extremely careful when removing and replacing ICs. All surface-mounted components should be replaced with components that have the exact part number. You need a pair of tweezers and a small-wattage pointed soldering iron for removing and replacing surface-mounted components.

10-4 The regular-size components are mounted on the top side of the chassis in this RCA 1993 portable TV.

10-5 Notice all of the black-looking SMD components on the bottom side of the etched wiring. Some SMD parts are mounted even on the center of a large IC that is mounted on the top side.

10-6 The pen points to a SMD transistor soldered on the PC wiring.

10-7 Besides TVs, SMD parts and IC processors are used in camcorders and VCRs.

Troubleshooting surface-mounted (SMD) components

Like any electronic circuit, take critical voltage and resistance measurements on surface-mounted parts. A fluorescent magnifying lamp and SMD component layout helps to locate the tiny defective part. Be extremely careful not to break any part or connection while working around these small components. Do not slide the TV chassis across the service bench or you can damage SMD components on the PC wiring side. Be careful not to short out circuits or surface-mounted devices (Fig. 10-8) when taking voltage measurements.

SMD parts have a tendency to break contact right where they are connected and soldered to the PC board. The large etched PC board can warp or flex enough to pull PC wiring connections. Remember that resistor and bypass capacitors look somewhat alike. In fact, some SMD parts might just be feed-through connections, which would show a shorted measurement on a DMM. Check these parts with the low ohmmeter range of a DMM. Recheck these small parts on an SMD parts layout, if available. Some resistors and capacitors are flat, and others are round with solid terminals. Use a fine-pointed soldering iron or a battery operated iron to solder the connecting ends of SMD components (Fig. 10-9). A small pair of tweezers are handy to place the miniature part in the right spot while soldering each terminal into the circuit.

Flyback PC ring mounting holes

The flyback or horizontal output transformer might be mounted directly on the PC board with ringed soldered eyelets. The pin terminals fit inside the ringed holes. To remove the defective flyback transformer, the excess solder must be removed on

10-8 Be careful when taking voltage measurements on SMD parts so that you don't short out the components and etched wiring.

all eyelets before the transformer can be pried from the PC board. If not, the board PC wiring or eyelet might crack and provide intermittent conditions.

When removing the excess solder with a 200- or 300-watt soldering gun and mesh solder wick, go over each eyelet three or four times to suck up the melted solder. Greater heat must be applied to the soldered terminal to remove all excess solder. Sometimes the small sucking type of iron will not clean up the soldered holes. After removing the transformer, clean up excess solder around the PC eyelet. Mount the new transformer and apply enough solder to fill each hole.

Brush and clean off all solder and rosin around the eyelets. Inspect the board for possibly poor connections. To be sure that the contacts are good, connect one ohmmeter probe to a flyback eyelet terminal and the other probe to where the same PC wiring connects in the circuit. Most PC wiring breaks occur right where it connects to the soldered eyelet.

Broken PC wiring

If a TV is dropped, inspect the board wiring with a magnifying glass. PC boards often break on sharp corners or where shields and heavy components are mounted. If the board is broken in several places, it's wise to replace the whole board.

Small breaks or cracked boards around parts can be repaired with hookup wire. Locate all broken PC wiring. Sometimes heavy components have a broken terminal or two. Scrape off the end of the PC wiring to be bonded together. Lacquer must be removed to make good contact. Scrape each broken end wiring with a sharp tool to get down to the copper PC wiring.

Do not try to solder the two broken ends together with just solder and the soldering iron. The crack will remain. Place hookup wire across the broken area and solder each end. If several components or bare, soldered joints are close by, use covered hookup wire. Doublecheck all connections along the cracked area. Be sure that each broken PC trace is on the exact wire. Bare hookup wire pieces can be used where it will not touch other connections. Now check each broken connection with the ohmmeter. Go from the first soldered connection to the next to ensure good bonds.

After the TV is operating, lightly push down around the cracked area with something nonconductive to see if the set has an intermittent connection. Likewise, push down on small components nearby to find a broken connection or part. Sometimes it is worth it to try to repair the board because it could take some time to obtain a new board.

Modular chassis design

Soon after the all transistor chassis was designed, the modular system was adopted by most every TV manufacturer. Modules are easily removed from the chassis and replaced with a new one (Fig. 10-10). The modules are constructed of lightweight components for use in practically any circuit. The large or heavy parts, such as transformers, filter capacitors, and picture tubes are mounted directly on the

10-9 Choose a small soldering iron to solder small SMD components.

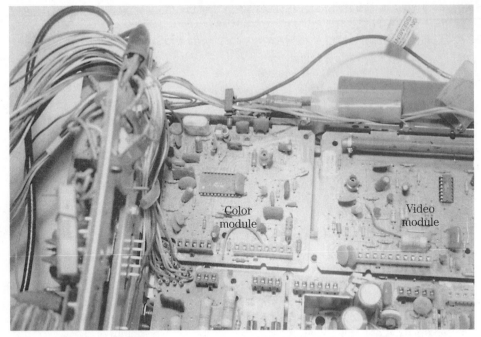

10-10 Choose the correct module from symptoms on the TV screen and replace it.

chassis. The modules can be plugged directly into a socket on the chassis or into a socket on a subassembly with wire leads connecting the module to the chassis.

Each TV manufacturer has its own set of modules. In other words, an RCA module will not interchange with a Zenith module and vice versa. However, some modules with the same circuit function can be interchanged with another chassis of the same manufacturer. Most modules are replaced instead of trying to repair or locate the defective component in the module. In some cases, more than one module has to be replaced to correct the problem.

In some brands, the vertical and horizontal circuits are located in one module, but in other TVs, the vertical and horizontal circuits are on separate modules. The video, brightness, and sync circuits are located in one module in most TVs. The modules might be mounted flat or on edge in the chassis.

The latest TV manufacturing trend is getting away from separate circuit modules, and you might find only the tuner, electronic tuning, and control circuit as separate modules in the TV. All other circuit components are soldered and mounted on one large PC board. Although, these late-model receivers cannot be classified as a modular TV, the control and tuner modules are plugged into one another and are connected to the TV with cables and plugs.

Module replacement

Before exchanging the various modules, you must know which modules to interchange. A schematic with a chassis module layout is essential. A lot of the TV manufacturers have drawings or photo layouts of the different modules. Look for a module location chart inside the TV cabinet. Practically anyone can interchange a module if they know what to look for and where to look. Some TV owners purchase a complete package of modules so that they can service their own TV (Fig. 10-11). Check the following service suggestions before attempting to replace a possible defective module.

Power-supply problems

SAW filter component

The surface acoustic wave (SAW) filter utilizes the surface wave of a piezoelectric element of very compact design. The SAW filter is a finger-like electrode that is arranged in length and width to ensure the desired bandwidth characteristic of the IF frequency. A surface acoustic wave filter device establishes the proper IF response. The SAW filter can be located after the tuner or between the tuner and IF amp and video detector (Fig. 10-12). The SAW filter is in a grounded, shielded container.

Very seldom does the SAW filter cause any service problems, but it can be checked with the DMM for leakage. Check each SAW filter terminal with the ground terminal for continuity. Any measurement between common ground and other elements indicates leakage. A normal filter has no continuity measurement. The unit might become cracked or open. You can check the SAW filter with a crystal checker in or out of the circuit (Fig. 10-13). Often, a lower measurement is found with in-circuit tests. A higher reading results when the SAW filter unit is out of the PC board. If the IF signal will not pass through the SAW filter device, suspect that the unit is defective.

10-11 The only modules used in recent TVs are the tuner and control modules.

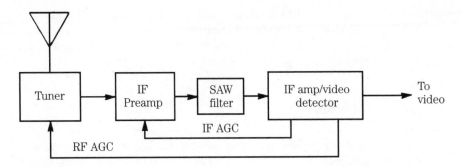

10-12 A block diagram of a saw filter network between the IF and IF amp.

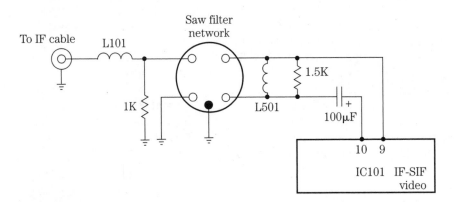

10-13 Here, the SAW filter is located between IF cable from tuner and IF/IF IC101.

Sand-castle generator

The sand-castle waveform is developed by combining the horizontal blanking, vertical blanking, and burst keying pulses applied to the luminance/chrominance processing IC (Fig. 10-14). The output of the horizontal oscillator is fed to the sand-castle generator. During the horizontal and vertical blanking period, these three pulses are combined to form the sand-castle waveform. The three-level signal waveform is fed from the sand-castle generator to the luminance/chroma IC. The sand-castle waveform can be observed at the luma/chroma IC input pin. The IC has an internal decoder network that decodes the three signals and sends them to the correct circuits. If the sand-castle generator is defective, there will not be any luminance or color from the luminance/chrominance processing IC.

Chopper power supply

The chopper power supply consists of a low-voltage bridge rectifier circuit providing voltage to the primary winding of the chopper transformer. The dc voltage from the chopper transformer is connected to the chopper transistor. Most chopper power supply circuits can be serviced with a DMM (Fig. 10-15).

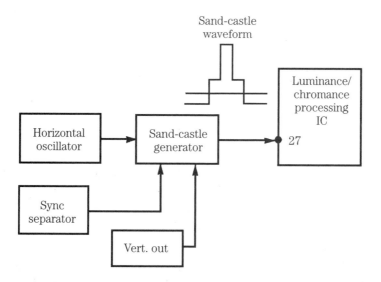

10-14 A block diagram of the sand-castle waveform generator.

10-15 Check the chopper output transistor and the voltages in the chopper power supply.

The pulse-width-modulated (PWM) chopper regulated power supply works like the horizontal deflection systems in many of the later TVs. The regulator control is a free-running oscillator with a frequency of about 15 kHz. This oscillator is triggered by a horizontal pulse derived from the secondary of the IHVT transformer, which locks the regulator to the scan frequency (Fig. 10-16).

The output of the PWM regulator circuit is applied to the regulator drive transformer. The output of the drive transformer is connected to the base terminal of the chopper output transistor. This turns the chopper transistor on and off. The raw dc voltage from the bridge rectifier is applied through the primary winding of the chopper transformer to the collector of the chopper output transistor (Fig. 10-17).

The on/off state of the chopper output circuit causes a pulsating dc voltage to occur in the primary winding of the chopper transformer. This induces pulses in the secondary winding of the chopper output transformer and is then rectified to provide voltage to several different voltage sources (Fig. 10-18).

Scan-derived voltages

Scan-derived voltages, used in many TVs, supply different dc voltage sources from the flyback transformer windings. The vertical, horizontal, video, sound, and luminance/chroma circuits are powered from the horizontal output transformer. Remember, when voltages are taken from the flyback, the horizontal circuits must be operating. Some chassis provide a voltage from the low-voltage bridge circuits to start up the horizontal circuits (Fig. 10-19), although other manufacturers have start-up circuits that eventually run after the horizontal circuits are started.

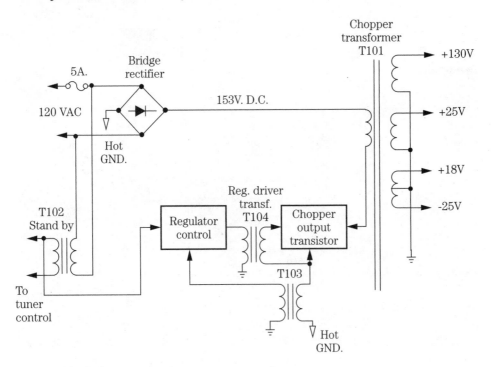

10-16 A block diagram of a chopper power supply.

10-17 The location of the chopper transistor on a heatsink in the RCA CTC132 chassis.

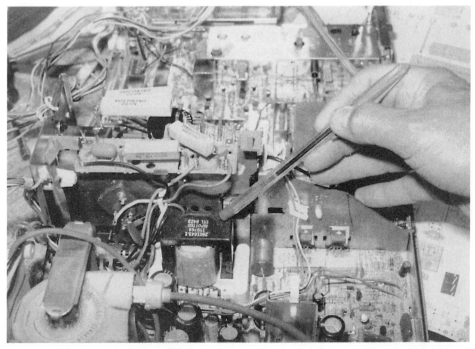

10-18 The chopper transformer is connected to the primary winding of the chopper transformer.

10-19 Scan-derived or secondary voltages are taken from integrated (IHVT) horizontal output transformer.

The scan-derived voltage sources are taken directly from additional windings on the horizontal or flyback transformer. Each winding is rectified by a small silicon diode and capacitance filter. Zener diodes can be used to provide additional regulated voltages (Fig. 10-20). Higher voltage sources for the color output transistor and picture-tube circuits are taken from another winding. When a leaky component within one of these secondary voltage sources occurs, the overloaded circuit might shut down the horizontal output transistor and all voltage sources. To determine if the flyback transformer or overload voltage source is defective, slowly raise the ac line voltage with an isolated power transformer.

Switched-mode power supply

The advantage of the switched-mode power (SMP) supply is to slowly bring up the rectified power-line voltage to prevent damage to critical ICs and transistors within the TV circuits. The SMP supply contains a self-oscillating transformer and voltage-regulator circuit (Fig. 10-21). The switched-mode power transformer operates from an ac-line-derived +155 Vdc. T402 also provides power-line isolation with the optoisolator (IC 403).

Magnetic energy is stored in the transformer during the on time and then is transferred to the secondary winding of switched-mode transformer during the off time. The amount of energy transferred is controlled by the switched-mode regulator transistor. The frequency of the power supply, under normal load, can range from 20 kHz to 40 kHz.

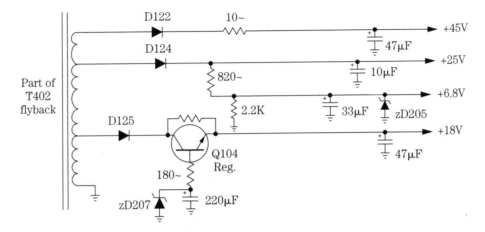

10-20 The secondary or scan-derived power-supply circuits taken from the flyback windings.

The differential amplifier (Q407) monitors the output voltage and compares this voltage to a zener diode. The connection voltage is fed to the main control amplifier transistor (Q404) and fed back through the optoisolator (IC403). The control circuitry varies the duty cycle and maintains correct voltage output.

Video problems

No sound and no raster

First, check for a possible defect in the circuit breaker or fuse. Replace the fuse or check for continuity with the low range of the ohmmeter. Now, look on the module layout chart for the location of the low-voltage module. Replace the low-voltage module with a new one. Plug in the TV and listen for the sound to come on. If you hear the sound, this often means the horizontal and output circuits are functioning normally.

If you don't hear the sound or see the raster, proceed to the horizontal oscillator and driver module. Interchange the horizontal modules. Most problems in any TV are caused by the horizontal circuits. A defective horizontal output transistor might not be physically part of the suspected horizontal module and might be located on the TV chassis. In other modules, this transistor is located in the horizontal module. A defective horizontal output transformer and defective high-voltage circuits can cause a no sound/no raster symptom.

Insufficient height or horizontal white line

A horizontal white line and insufficient vertical height indicates problems within the vertical circuits. Some modules have all the vertical components mounted on one module. In others, the vertical output transformers are mounted on the chassis. Usually, replacing the vertical module eliminates the vertical height problem.

If the raster almost fills out the screen after replacing the vertical module, readjust the vertical size control. Just one vertical control might be located in the

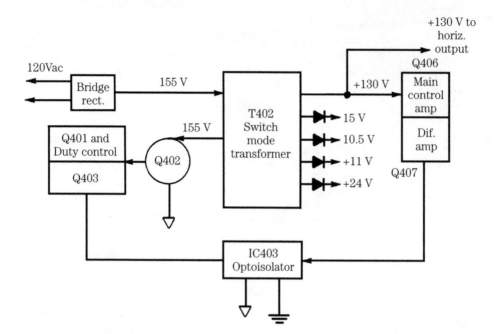

10-21 A block diagram of the switch-mode power-supply circuits.

transistor chassis. The raster should completely fill out the screen after the adjustment. If not, other components could be defective, such as an open vertical coupling capacitor between the vertical circuit or the yoke might be open. The capacitor values might range between 470 μF and 4700 μF. A horizontal white line might also be caused by an open yoke winding or a poor vertical yoke socket.

No raster and normal HV

Suspect a defective video output circuit or the voltages applied to the picture tube to cause a no raster problem. You can detect high voltage in the picture tube by feeling the front of the TV tube or listening for the sound of the yoke expanding when you first turn the set on. Place your arm next to the TV; if a high-voltage potential is present, you will feel the hairs on your arm being attracted toward the screen. If there are no signs of high voltage, check the filament of the tube and see if it is lit.

Now, replace the video-luminance module. You can find a separate matrix-color amplifier module or the luminance driver transistors are mounted on the CRT (cathode-ray tube) socket. Replace both the video luminance and video drive modules, if necessary, to achieve correct brightness. In some chassis, a kinnie module might be used for each color output circuit. Excessive or insufficient brightness can often originate in the matrix-power amplifier stages.

Snowy or weak picture

A snowy picture can be caused by a defective tuner, IF stage, or outside antenna. Cleaning the tuner might solve a snowy or erratic picture and sound symptom. Replacing the tuner module in the latest color chassis can cure a snowy or weak picture. A snowy picture can be caused by a defective AGC circuit.

Check and see if there is one module includes the entire IF and AGC circuits. In some TVs, the sync and AGC circuits are located in the IF module. In other TVs, the AGC and sync circuits are located in a separate module. Replace the IF and AGC modules after replacing the tuner or checking the tuner for possible defects.

Poor sync

Suspect a defective sync circuit if the picture drifts either vertically or horizontally and cannot be adjusted with either corresponding control. Notice if these sync circuits are located separately or are a part of the IF module. Sometimes, just replacing the IF module will cure both the out-of-sync and snowy picture. Always adjust the AGC control after replacing the AGC module. The AGC control might not exist in some of the latest TVs.

Color problem

Color problems originate in the color IF, demodulator stages, and color video-output circuits. Just replacing one color module might not solve the weak or no-color problem. You might find two different color modules in the color circuit of one color chassis and all of the color circuits on one large PC module in another color chassis (Fig. 10-22).

Before attempting to replace the color modules, adjust the fine-tuning control. Tune in the best possible picture and sound from a local TV station. A defective or

10-22 Some older TVs have two or three color modules instead of one.

misadjusted color-killer control can cause color rainbows or missing color. Check the manufacturer's literature for the correct color-killer adjustment. Replace the color module if no color-killer control is on the TV.

Sound problems

In the early solid-state TV, the sound circuits used transistors. Today, one large IC might contain all audio output sound circuits (Fig. 10-23). Sometimes the sound circuits are included in one large IC that contains luminance/chroma, video, sync, AGC, and deflection circuits. You might find a combination of sound IF and discriminator circuits in one IC and the audio output circuits with transistors.

Often, the sound takes off after the first video stage to a sound IF transformer. The audio IC processor amplifies the weak audio signal and the discriminator coil tunes in the FM audio signal. The audio signal is controlled by the volume control and passed on to the audio output stage, and eventually the speaker. A slight touch-up of the discriminator coil might clear up the muffled or distorted sound if the sound signal is off frequency (Fig. 10-24).

The audio signal can be signal-traced with a signal tracer or external audio amp from sound detector to the speaker. Loud popping sounds might be caused by the large audio output transistor. Intermittent sound might be caused by a defective IC or electrolytic coupling capacitor. Weak sound might result from a leaky or open transistor, open coupling capacitors, or a dried-up electrolytic speaker coupling capacitor.

Stereo audio

Most of the higher-priced TVs have stereo sound. Today, some TV stations are broadcasting stereo audio, which is signaled by a red LED. Not all TV programs are broadcast in stereo. When stereo sound is broadcast, the audio comes out of two separate amplifiers and speakers.

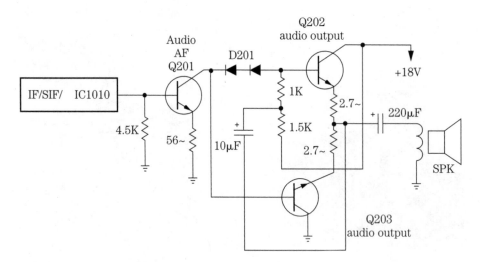

10-23 You might find three transistors in the sound-output circuits of the latest TV portable.

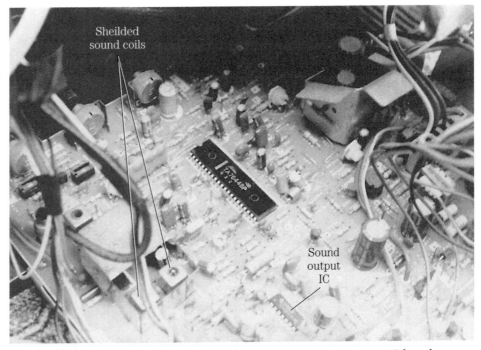

Sheilded
sound coils

Sound
output
IC

10-24 Locate the shielded coils to locate the audio output circuits on PC board.

Within the stereo sound circuits, the stereo or mono sound signals are applied to the stereo demodulator circuits (Fig. 10-25). The baseband audio signal is developed by the system-control IC, which determines when a stereo signal is received. This stereo signal is passed on to the stereo demodulator (IC1). The stereo audio signal is applied to a matrix and differential amp IC (IC2). The outputs of the differential amp and matrix are separate left and right audio channels. The L and R audio signal is switched by IC3. The switching IC applies the mono or stereo signal to the audio amplifiers.

The stereo or mono signal passes on to volume control IC4. Here, the signal controls the volume and can be muted by manual or remote control (Fig. 10-26). The left and right audio signal is applied to the dual audio IC amp (IC5). Here, the weak stereo audio signal is amplified and coupled to the stereo speakers with electrolytic coupling capacitors. The audio stereo signal can be signal-traced like any amplifier, starting at the audio sound output of the matrix (IC2).

Weak or no sound
Replacement of the sound module can cure all sound problems. When modules first were used in the black-and-white or color chassis, the sound IF and pre-amp stages were mounted as a plug-in module. The sound output transistor was located on the chassis. In the modular chassis, the whole sound circuit might consist of one IC with its associated components located on one module.

If, after replacing the sound module, the sound is still weak, suspect an open or defective (speaker to module) electrolytic coupling capacitor. If defective, these small capacitors might produce weak, intermittent, and distorted sound. They are

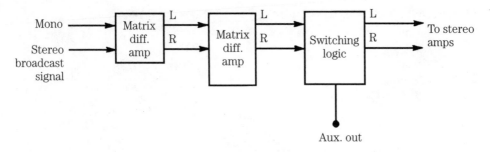

10-25 A block diagram of the stereo sound circuits in a recent TV.

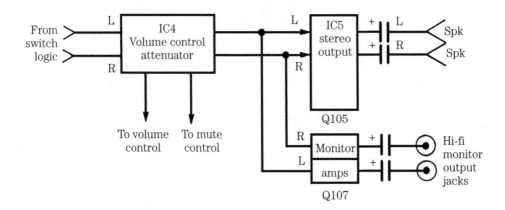

10-26 A block diagram of the stereo audio output circuits with stereo output jacks.

easy to locate by tracing the speaker wiring to the PC board and the PC wiring to the capacitor. The coupling capacitor is located between the output transistor and the speaker cable wires.

Be careful when replacing any TV module. You can easily crack the PC board by improper mounting or with too much pressure. Be sure that the module is properly seated. Doublecheck all external plugs connecting the chassis to the module. Check for any bent plugs or socket pins before mounting module. If several wire plugs are removed from the module before replacement, each plug should be marked so that they can be properly connected. Do not clean the module contacts with a liquid cleaning solution. Clean off all the module contacts with a pencil eraser. A quick modular troubleshooting chart is found in Table 10-1. Replacing a suspected defective module can quickly solve the dead or erratic TV problem and save a few service dollars. You might want to leave module replacement up to the professional technician.

Parts layout

Before you can find or replace a suspected or defective component, the part must be located in the TV. Some parts are easy to locate, such as power transformers, flyback transformers, and the picture tube. Others must be pointed out in a parts layout chart. Most manufacturers have a parts-layout chart with photos and drawings labeling the different components. Sometimes critical part layout charts are glued to the side of the TV cabinet. Transistors can be located on a cabinet chart. Of course, the TV technician has original manufacturer's schematic and layout charts for easy reference.

Other components

Besides transistors, the TV might also have special diodes, IC circuits, SCRs, and tripler units. In many of the vacuum-tube chassis, silicon diodes were inserted in the low-voltage rectifier circuit. Today, special switching, zener, and damper diodes are used throughout a the TV. The silicon-controlled rectifier (SCR) can be used as a regulator in both the low- and high-voltage circuits. The latest color TV sets have high-voltage diodes molded right into the flyback transformer, eliminating the need for separate regular tripler units (Fig. 10-27).

The integrated high-voltage transformer (IHVT) has many different windings, compared to the regular flyback or horizontal output transformer. The high-voltage

Table 10-1. A quick modular troubleshooting chart.

No sound and no raster:	Check for open fuse Reset circuit breaker Replace low voltage module Replace horizontal module
Sound and no raster:	Check for light of CRT heater No high voltage Replace video/luminance module Replace matrix and color output module
Snow picture:	Clean tuner Check antenna signal Replace IF AFT module Replace A AGC, IF, and RF module
Poor horizontal sync:	Adjust AGC control Replace AGC-sync module Replace horizontal oscillator module
Poor height and linearity:	Adjust vertical height or size control Replace vertical module Poor vertical sync: Adjust AGC control Replace AGC-sync module
Poor or weak color:	Check fine tuning and color-killer control Replace chroma module Replace matrix-luminance module
Weak or distorted sound:	Replace sound module Replace IF module Check speaker-to-circuit coupling capacitor Check for defective speaker

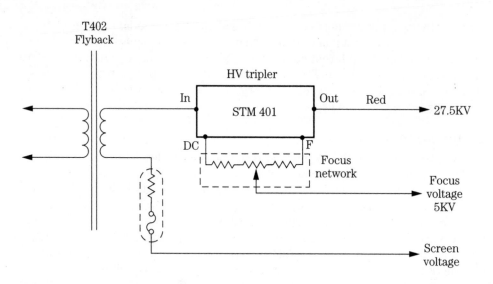

10-27 The tripler unit steps up the voltage of the secondary winding from the flyback to the required high voltage for the CRT.

winding that went to the high-voltage rectifier or tripler unit now has silicon diodes molded right inside the plastic molded case. A large high-voltage lead comes out of the flyback and goes directly to the anode connection of the CRT (Fig. 10-28), in addition to the primary windings that supply voltage sources to other TV circuits. Like the old flyback transformer, many of them become leaky, shorting HV diodes from the internal arcover, mandating replacement.

Before the use of IHVT transformers, tripler units were used that contain high-voltage diodes and capacitors to build up high voltage from the flyback transformer. They, too, can break down and arc over. Sometimes the HV arcs over from the tripler body to the metal TV chassis. The tripler unit has input, ground, and focus terminals. The large HV-insulated lead goes directly to the anode of the CRT.

Transistors are available in many sizes and shapes. High-powered transistors are used in the sound output, horizontal and vertical output circuits, and low-voltage regulator circuits. Medium-powered transistors are used in the vertical output stage, and in the sound stages. Low-powered transistors are used in the remaining solid-state circuits (Fig. 10-29).

The integrated circuit (IC) was first located in the sound preamp and driver stages of the tube chassis. Now, one IC can combine the entire audio circuit. Some ICs were located in the color stages of the tube chassis, but now the entire color, luminance, and video circuits are contained in only one large IC. A separate IC package might house the oscillator and driver stages of either the vertical or the horizontal circuit or both of the circuits can be included in one IC (Fig. 10-30). Replacing just one IC can solve several problems in several different circuits.

Low-powered signal diodes are located in the RF, IF, and video circuits of the TV. In contrast, the damper diode is a special heavy-duty type with a very high voltage rating and is used in the collector circuit of the horizontal-output transistor. Always,

IHVT
transformer

10-28 The integrated horizontal output transformer provides several low-voltage sources besides the HV, focus, and screen voltages.

Horiz.
output
transistor

10-29 The transistor on a heatsink in the horizontal output circuits.

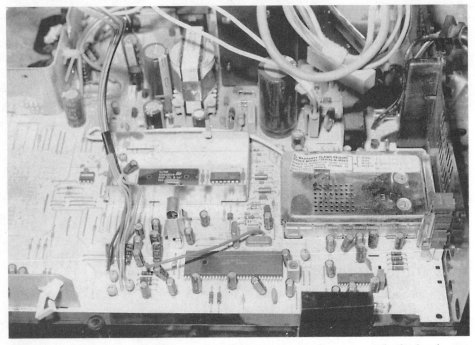

10-30 Several different ICs were used in older TVs, but today, you might find only two large ICs in the entire chassis.

be careful to observe polarity when installing a new damper diode and be sure that it has the correct voltage rating. Zener diodes are used in the low-voltage circuits. They are used to start, stop, and shut down the regulated voltage circuits. The zener diode has a tendency to become leaky.

Checking the TV chassis without a schematic

Many defective components in the TV can be located by sight, sound, and smell. Visual inspection and touching components have repaired thousands of TVs. Check for burned boards, resistors, arced-over parts, cracked boards, overheated components, and general wear and tear (Fig. 10-31). Arcing between parts can be caused by liquid spilled down into the back area of TV. Smoke from a flyback or yoke assembly might indicate that the windings are shorted. Cracks in ceramic capacitors and resistors indicate components have overheated. Check for a leaky transistor, diode, or IC nearby.

You can hear high-voltage arc-over and cracking noises around the flyback or picture tube, indicating excessive high voltage or leaky HV diodes (molded inside the horizontal output transformer). Hold-down (safety) capacitors in the horizontal output circuits can open, creating excessive high voltage. A tic-tic noise in the flyback might indicate chassis shutdown with a defective deflection or flyback circuit.

Hum in the speaker can be caused by a defective or open filter capacitor in the low-voltage power supply. Hum and distortion can result from a leaky output IC or transistor. The mushy sound in the speaker can be caused by a dropped or frozen cone.A vibrating ferrite bead can produce a high-pitched sound on the chassis. Loose particles within the flyback or transformer mounting frame can cause a vibration noise.

The burned power transformer or flyback can produce an odor, indicating burned windings or wiring. Besides smelling hot, if you place your fingers near the horizontal output transistor, you can tell if it is operating too hot. Check for a leaky transistor or an improper drive voltage. Improper drive voltage, leaky yoke, flyback, and overloaded secondary voltage circuits can cause overheated output transistors. Overheated surge or isolation large-wattage resistors in the low-voltage power supply indicates leaky components in the voltage source or overloaded connecting circuits. If you cannot see the CRT light, you can feel the glass neck of the tube to determine if it is warm and operating. Smoke or smoldering PC wiring indicates arcing between components on the PC board.

Look up the transistor or IC part numbers found on the body of the defective component and check in the universal solid-state replacement manual to what circuit they belong in, the correct terminals, and the working voltage. Not only does the part number tell you what circuit it works in, but the correct replacement. If the correct schematic is not available, look up one that is similar to the same chassis number, because most manufacturers use the same circuits year after year.

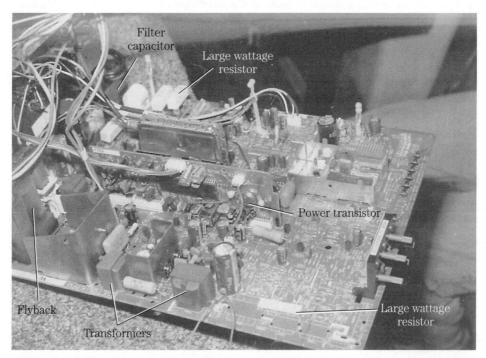

10-31 Check the chassis for burned marks, smoking flyback, overheated transistors, and burned high-wattage resistors.

High-voltage shutdown

High-voltage shutdown can occur if the high voltage reaches the X-ray protection circuit and shuts the chassis down. Excessively high voltage can occur if the ac power-line voltage raises above the 120-Vac range. Power-line voltages above 125 or 128 Vac might cause the chassis to shut down. An open or a change in capacitance of the bypass capacitor across the collector terminal of the horizontal output transistor can increase the high voltage (Fig. 10-32). A defective IHVT transformer might cause high-voltage shutdown.

To determine if the chassis is shutting down from excessive HV, monitor the HV at the picture tube, the waveform at the base of the horizontal output transistor, and the dc supply voltage with the DMM. Plug the TV into a variable isolation power-line transformer and slowly increase the line voltage from 60 to 80 volts. Notice how the high voltage is measured at 80 Vac. Advance the transformer until the chassis shuts down. If excessively high voltage is noted before the 120-Vac voltage is reached, check the horizontal output and flyback for defective circuits.

X-ray protection circuits

The X-ray circuit protects the operator and chassis from the excessively high voltage applied to the picture tube. The shut-down circuit operates when the voltage or current has reached a certain limit and shuts down the horizontal output transformer and chassis. In the early chassis, when the high voltage went over the manufacturer's limit, the protection circuits would turn the horizontal oscillator stages off frequency. When high voltage was present, you could not successfully adjust the horizontal control and bring the picture into sync (Fig. 10-33).

A pulse from the flyback winding is applied to the X-ray shut-down circuits and applies a rectified voltage to the shut-down circuits. As long as this voltage stays below the zener diode voltage, the diode will not conduct. When excessively high voltage is applied to the flyback, the pulse is larger and applies a higher voltage to the zener diodes, causing the diode to conduct. When the diode conducts, an SCR or diode provides bias applied to the horizontal oscillator circuits. The horizontal oscil-

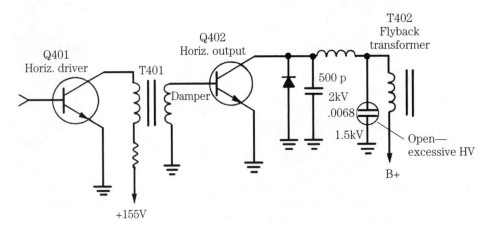

10-32 The voltage can become too high if the safety capacitor is open.

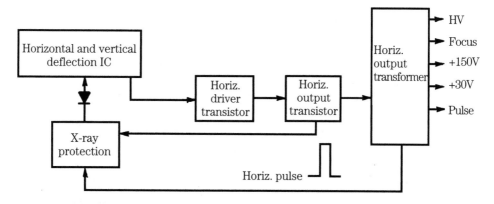

10-33 A block diagram of the x-ray protection circuit to shut down the TV chassis with excessive high voltage.

lator is disabled, and the whole chassis shuts down. Check for a pulse waveform from the flyback transformer with a scope. Correct voltage and resistance measurements might reveal a defective shut-down circuit.

Power-supply regulator circuits

In many of the lower-priced ac TVs, the power supply consists of a half-wave or bridge rectifier and an output regulator. The ac voltage is fuse-protected (F101) and is applied to the isolation resistor and silicon half-wave rectifier (Fig. 10-34). Again, the dc voltage is filtered with C1 and is fuse-protected with F102. The high dc voltage is applied to the input terminal of the low-voltage regulator.

The low-voltage rectifier might consist of an IC circuit that looks like a regulated horizontal output transistor, except that it has three terminals instead of two, besides the metal body of the regulator. Often, the metal body is insulated away from the metal chassis and is the regulator output terminal. In other TVs, the IC regulator resembles a power output IC used in high-powered stereo audio-output circuits. You might find low-voltage regulators of 115, 120, 125, and 130 volts in the ac TV. The correct voltage regulator must be replaced with the exact replacement or a compatible universal part. The defective regulator might short, or become leaky or open.

Standby power circuits

The standby power supply provides voltage at all times to the IR receivers and start-up circuits of the horizontal output transformer. Ac power is applied to the standby transformer (T401). The low ac voltage is rectified and applied to several standby-voltage sources. Often, the standby power supply is responsible for standby voltage and regulation (Fig. 10-35).

In this circuit, 21 volts is applied to the horizontal driver transistor during start-up. The 11.5-volt source supplies voltage to the infrared receiver and keyboard. A 5-volt standby source is fed to the system-control IC. The standby voltages are needed when the remote or keyboard is pressed to activate the TV. Notice that these standby circuits are on all the time. Of course, until the chassis is activated,

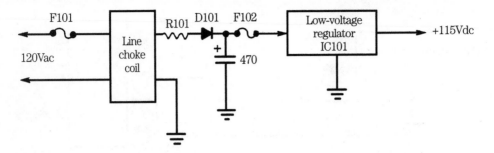

10-34 A block diagram of the fixed low-voltage power regulator IC.

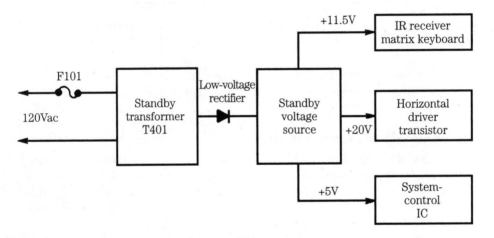

10-35 A block diagram of the standby circuits in the remote-control TV chassis.

very little current is being used. Check the standby circuits if the remote will not turn on the TV, provided that the remote transmitter is functioning.

Horizontal and vertical countdown circuits

The horizontal and vertical countdown circuit can be controlled with a fixed-crystal oscillator. The horizontal and vertical oscillator circuits are located inside of the processor IC. The frequency of the horizontal oscillator is determined by the crystal. In the RCA chassis, the output of the crystal is 32 times the horizontal frequency.

The crystal operates at 503 kHz. The output of the VCO is applied to the horizontal countdown circuit (Fig. 10-36). The countdown circuit divides the crystal frequency to 15,734 kHz. The output of the horizontal countdown circuit is amplified and a square-wave pulse is applied to the drive circuits.

The vertical circuits can operate from a countdown divider circuit. The vertical sync and separator circuits lock in the vertical countdown circuits. The 15,734 kHz frequency of the horizontal oscillator is applied to the vertical up/down counter and horizontal frequency doubler circuitry (Fig. 10-37). The output of the vertical drive (sawtooth waveform) is connected to the vertical output circuits, which scan the vertical deflection winding.

IC deflection circuits

When the solid-state TV was first designed, only transistors were used in the vertical and horizontal deflection circuits. Today, the horizontal and vertical oscillator circuits are located in one large IC. Transistors or ICs can be used in the vertical and horizontal driver circuits. The horizontal output transistor is used in most TVs, but the vertical output circuits can be transistors or ICs (Fig. 10-38).

In this vertical deflection system, the deflection (IC1001) pulse is applied to a vertical reset and sawtooth transistors. The error-amp transistor provides a pulse to the input of the vertical output (IC103). Output terminal 7 applies a 55-volt p-p waveform to the vertical yoke winding.

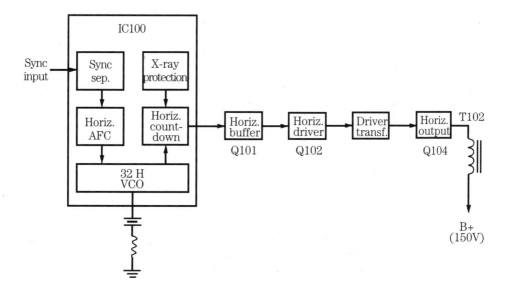

10-36 A block diagram of the horizontal countdown deflection circuits.

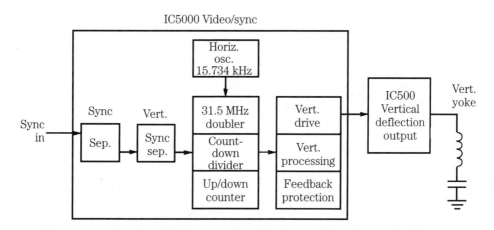

10-37 A block diagram of the vertical countdown deflection circuits.

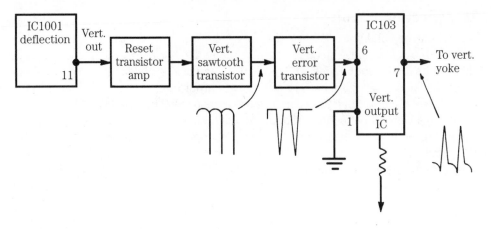

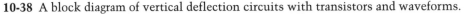

10-38 A block diagram of vertical deflection circuits with transistors and waveforms.

The horizontal deflection IC and amp supplies a 10-volt p-p positive pulse to the horizontal driver transistor (Q401). The collector of the driver transistor applies a pulse to the horizontal driver transistor (Fig. 10-39). The base of the horizontal output transistor is tied to the driver transformer. The collector terminal of the horizontal output transistor is tied into the primary winding of the yoke. Notice that the damper diode is inside of the horizontal output transistor. The regular horizontal output transistor will not function in this circuit.

System control

The system-control circuits in the latest TVs are controlled with a large IC processor or microcomputer, like those found in CD players and camcorders. The system-control microcomputer generates the clock signal and data. The information transmitted on the serial data bus includes local and remote commands, AIU utilization data, and tuning commands (Fig. 10-40).

The system-control microcomputer contains random-access memory (RAM) and read-only memory (ROM). The RAM portion stores information that frequently changes (such as channel scan) and customer settings (such as picture and volume-control adjustment). The AIU processor allows the set to return to the last channel viewed with the same volume and picture adjustments—even after the set is turned off.

The ROM portion of memory is used to store information that never changes, such as programming for the microcomputer, on-screen display, tuning, etc. Some manufacturers call the system control the *tuning system*. The 5-volt power from the standby power supply keeps the microcomputer "alive" during power-off periods.

The tuning section selects channels by remote control, channel scan, and random access from the remote transmitter. The on-screen display and menus contain the audio, color, and picture, clock, and channel adjustments. Each menu can be selected by pressing a menu button on the remote control.

The system control can be serviced by determining what signals are present on an oscilloscope. The clock oscillator frequency can be checked with the scope. Critical voltage measurements on each IC or microcomputer terminal can indicate that a circuit or IC is defective. The system-control circuits are quite complex and should be serviced by the electronic technician.

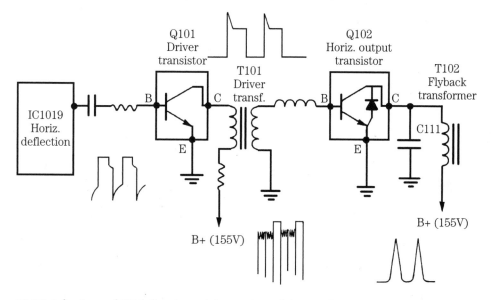

10-39 A horizontal IC deflection with transistor driver and output transistor circuits.

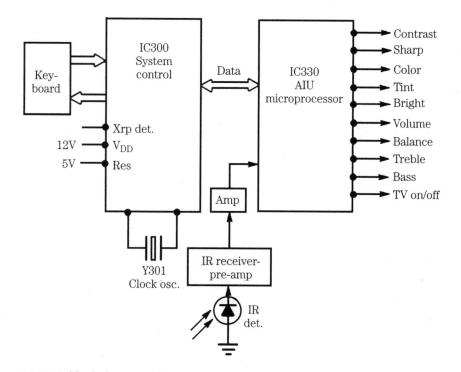

10-40 A block diagram of system-control tuning with IC and microcomputer operation.

11
CHAPTER

Solid-state servicing

Transistors and ICs are utilized in every circuit in the TV chassis. First, the transistor was used in the sound circuit of the TV. Next, the transistor was placed in the tuner, IF, and sync circuits of the hybrid chassis. Later, as more powerful transistors were developed, they were installed in the horizontal output and regulator circuits, resulting in the all-transistor chassis (Fig. 11-1).

11-1 The TV from 1996 is quite small compared to that of 20 years ago.

Both npn and pnp transistors are utilized in the circuits of a solid-state chassis. Although, in the older TV, a greater number of pnp germanium-junction transistors were used, that trend has reversed and now more npn transistors are used in the circuits of color receivers. The npn silicon transistor has a positive voltage applied to the collector terminal. The reverse is true of the pnp transistor, and a negative voltage is connected to its collector (Fig. 11-2).

A defective transistor can be located with an accurate voltage or resistance measurement of the device in the circuit. If the suspected transistor is removed, the transistor can be checked with a regular transistor tester or ohmmeter. The suspected transistor can be checked in or out of the circuit with a digital multimeter transistor diode tester. The suspected IC can be located with accurate voltage and resistance measurements and "signal in versus signal out" tests.

The beginner can make many tests and adjustments to a TV. You can test many components with the transistor tester and DMM (Fig. 11-3). Besides taking voltage and resistance measurements, you can check resistors, capacitors, coils, diodes, and transistors with the DMM. Careful voltage measurements on ICs might indicate that an IC is defective. You can purchase a portable DMM from $20 to $99. Pick up a low-priced DMM and give it a try.

Many different tests and adjustments for the beginner are listed in this chapter along with several tests for the more-experienced person. The intermediate or advanced person can often have acquired some experience from home or workshop facilities and is able to operate some test instruments. Of course, the beginner should only go as far as he or she feels confident to go. A beginner should not attempt to make extensive repairs. It is better to call in a professional electronic technician.

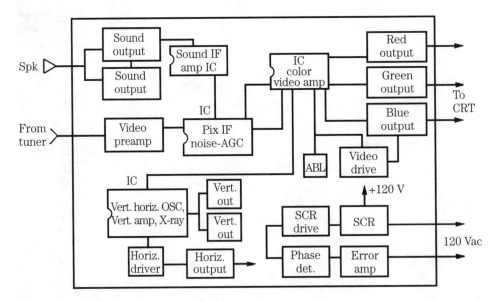

11-2 Today, TVs are filled with transistors and ICs.

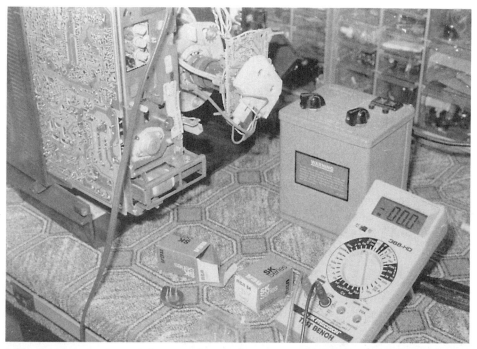

11-3 Critical transistor, voltage, and resistance tests can be made with the DMM.

Critical voltage and resistance measurements

Critical voltage and resistance measurements on a transistor or IC can indicate a leaky, open, or shorted part. You will find very low voltage measurements of less than 1 volt within the chassis. The forward-voltage measurement between the emitter and base is 0.6 volts in a silicon (npn) transistor. Low-voltage measurements on the base and emitter terminals to ground are normally low and can be accurately measured with a digital multimeter (DMM).

Low-resistance measurements of emitter bias resistors can be less than 1 Ω. A 0.5-Ω resistance can be measured on some audio-output emitter resistors. Continuity of coils, transformers, and speakers have a very low resistance. Critical resistance measurements can determine if a suspected IC is leaky. High-resistance measurements above one megohms can indicate that high-resistance resistors have changed value. Always remove one end of a resistor and diode from the circuit for a correct resistance measurement. Choose a good DMM for accurate voltage and resistance measurements.

Transistor resistance tests

Many transistor testers are on the market to help locate the defective transistor. Most transistors are checked for an open, leaky, or intermittent condition. These transistor testers can check the transistor in or out of the circuit. Test for an open base to collector and base to emitter. Testing can be done with an ohmmeter. In-circuit leakage tests within the circuit could result in an erroneous measurement from diodes, coils, or low-bias resistors being connected across the transistor terminals. The transistor might test normal in or out of the circuit, then break down under the circuit load and become intermittent. An intermittent transistor is one of the most difficult components to locate. Replacing the intermittent transistor with a new one is best (Fig. 11-4).

The transistor can be checked with a pocket VOM in or out of the circuit with the low ohmmeter range. Remove the transistor from the circuit when a low resistance is measured between any two terminals. These resistance measurements should be made with the 2000-Ω scale, then repeated on a lower scale if a short or leakage is detected. After removing the suspected transistor, check the resistance between any two terminals. Actual resistance measurements of a leaky transistor are shown in Fig. 11-5.

Besides providing very accurate voltage and resistance measurements, some DMM have a diode and transistor test capability so that can quickly test all transistors in just a few minutes. Each transistor can be checked for an open, leaky, or normal condition right in the circuit. The digital multimeter (DMM) is a handy test instrument for checking a transistor or diode in or out of the circuit.

Rotate the function switch to the diode or transistor test position of the DMM. With npn-type transistors, place the red probe (+) to the base terminal. Now place

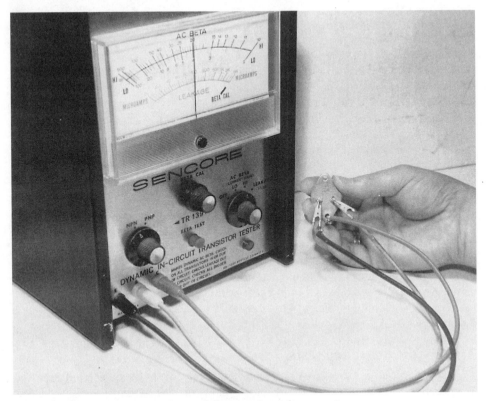

11-4 Check transistors in and out of the circuit with a transistor tester.

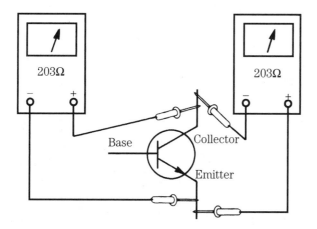

11-5 Low-resistance measurements between any two transistor elements in both directions indicates that a transistor is leaky.

the black probe (−) to the collector terminal. No reading will be indicated for an open junction between the base and collector terminals. A very low resistance measurement might indicate that the transistor is leaky. Reverse the test leads and check for a low-resistance reading in the other direction. Suspect that the transistor or diode is leaky when the meter indicates low resistance between the base and collector terminals. Likewise, place the black probe to the emitter terminal. A normal resistance measurement, comparable to the emitter or collector terminal, indicates that transistor is good (Fig. 11-6).

Now, check for leakage between any three terminal elements. Remove the transistor for out-of-circuit tests whenever you locate a low resistance between any two terminals. Suspect that a transistor is defective if it has a low-resistance between two terminals, but it tests normal after it's removed from the circuit. Double check the resistance measurement with the DMM switched to the 2000-Ω range.

The pnp transistor can be checked in the very same manner. Place the black probe (−) to the base terminal. You can quickly check the polarity of any transistor with the base terminal probe in relation to the emitter and collector terminals. The transistor is a pnp type if you measure a low resistance with the black or negative probe of the DMM attached to the base terminal of the transistor while the red probe is attached to the emitter or collector terminal. Likewise, the unknown transistor is an npn type if, in order to get similar results, you have to attach the red or positive probe of the DMM to the base terminal of the transistor (Fig. 11-7).

The zener diode can be checked for open, leaky, or normal conditions with the diode scale of the DMM. For a quick normal test, place the red probe (+) to the positive terminal of the diode (anode) and the black probe (−) to the cathode lead. The normal diode will show a low resistance measurement. A shorted or leaky diode will have a low resistance measurement with reverse test leads (both directions) (Fig. 11-8). Doublecheck the resistance measurement by removing one terminal of the diode. The open diode will not show a reading in any direction. Most defective diodes

11-6 A commercial transistor tester.

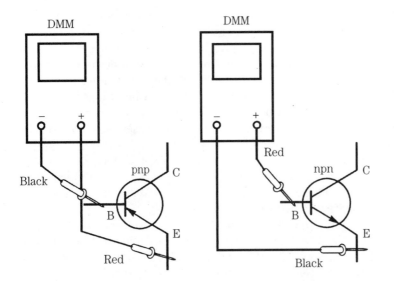

11-7 For npn-type transistors, place the red probe or positive lead on the base terminal. In pnp types, place the black probe on the base terminal for correct transistor tests with the DMM.

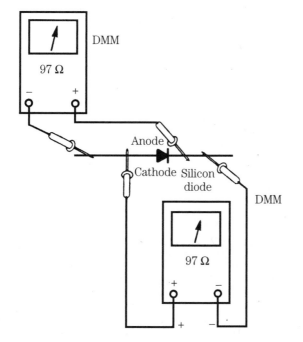

11-8 A shorted or leaky transistor and diode will have a low-resistance reading in both directions.

are shorted or leaky. The damper, boost, or high-voltage rectifiers might not provide any indication on the diode test mode of the DMM.

Transistor voltage tests

The defective transistor can be located by taking accurate voltage measurements on each terminal. Accurate voltage readings can be taken with the VOM or VTVM. The low-priced DMM is a very accurate voltage and resistance test instrument. First, take voltage, then resistance measurements.

A higher-than-normal voltage on the collector terminal might indicate that the transistor is open. Always obtain a schematic diagram when taking voltage and resistance measurements. Check for no voltage or low voltage on the collector terminal. No voltage on the emitter terminal might also indicate that the transistor is open (Fig. 11-9). Check the emitter bias resistor for correct resistance. An open emitter resistor might cause no voltage at the emitter and very high voltage at the collector terminal.

Low collector voltage might indicate that there is a leaky transistor, increased collector resistance, or improper supply voltage. Now check the emitter and base voltages. Suspect that a transistor is leaky if all three terminal voltages are about the same (Fig. 11-10). You might find some normal TV circuits, where the voltages are quite close on all three terminals. A lower collector voltage with a higher emitter voltage might indicate that there is leakage between the collector and the emitter elements. Make transistor in-circuit tests or remove the transistor to test it out of circuit

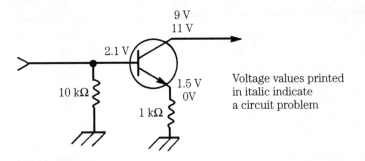

Voltage values printed
in italic indicate
a circuit problem

11-9 If no voltage is measured on the emitter terminal, the transistor might be open.

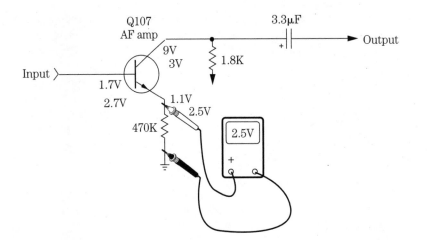

11-10 Suspect that a transistor is leaky if the base, emitter, and collector voltages measure about the same.

for leakage. Low collector voltage and a higher base voltages might indicate that there is leakage between the collector and base terminals.

Low-resistance measurements between any two elements might indicate that a transistor is leaky. Be sure that outside components (such as diodes, directly coupled transistors, coils, and resistors) are not making the low-resistance path (Fig. 11-11). Measure the base and emitter resistors for correct resistance. Disconnect one of the two transistor terminals when low resistance or leakage is indicated. Desolder it with a solder-wick or a special solder remover. Now, take another resistance measurement. Suspect that a transistor is leaky if a low-resistance is still present.

Critical resistance and voltage measurements might be too much for the beginner, but easy for someone with intermediate or advanced troubleshooting skills. The beginner can quickly learn electronics by taking voltage and resistance measurements of certain TV circuits. Voltage measurement on the high-voltage circuits should only be made by the advanced or professional electronic technician.

Checking integrated circuits

The IC plays a large role in the operation of TVs today. In the future, the entire TV chassis might only contain IC chips with only a few discrete transistors. With digitally operated circuits just around the corner, the TV will be completely controlled by ICs. The IC is a single device whose resistors, capacitors, diodes, transistors, and connecting wires are all made by processing a piece (chip) of semiconductor material in such a manner as to produce a complex circuit (Fig. 11-12).

To locate a defective IC, first find out which circuit is malfunctioning, then compare the malfunction with the purpose of each IC. Sometimes more than one integrated circuit is involved. Most failures can be located by tracing the signal in and out of the IC. Critical voltage and resistance measurements at the IC pins help you to indicate a defective IC circuit. Always have the schematic handy when trying to locate a defective IC.

The beginner might be able to locate a defective sound, AGC, or sync IC with voltage tests or with an IC replacement. Spray the suspected audio IC with a coolant if the sound is distorted, intermittent, or dead. Replace the defective IC if the sound appears normal after it is sprayed. Take voltage and resistance measurements at the IC terminals to verify whether or not the IC is defective. The AGC and sync ICs can be serviced in the same manner.

The more advanced technician might be able to service the IF, horizontal and vertical, and video ICs by injecting a signal in and observing the output with an oscilloscope. No vertical output pulse from the vertical oscillator or drive section can be caused by a defective IC. Improper horizontal oscillator signals can be caused by the

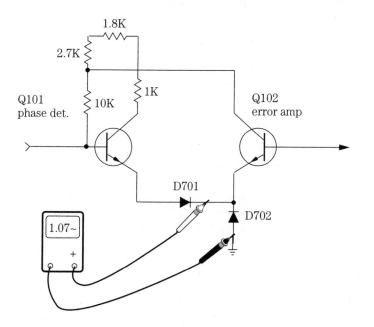

11-11 A low-resistance reading in the transistor circuits might indicate that a diode, transistor, or coil across the transistor elements is leaky.

11-12 ICs can have from 8 to 80 pin terminals.

same IC. Before removing the suspected IC, take accurate voltage and resistance measurements with a DMM. In many cases, a leaky capacitor or a change in the resistance connected to an IC terminal can cause the IC circuit to not function properly.

Typical sound IC service procedure

A dead audio section should be checked by clipping a new speaker across the TV speaker terminals. Turn the volume up halfway. Replace the defective speaker if sound is heard from the known-good speaker. If the sound is still dead, clip a 100-µF 25-volt electrolytic capacitor in series with the subspeaker (Fig. 11-13). In this circuit, attach the clip lead at pin 2 of IC201, if the sound returns, replace the open speaker capacitor. Many electrolytic coupling capacitors become open after a few years of operation. These two sound tests can easily be made by the beginner in electronics.

For weaker or distorted sound conditions, spray the IC with coolant. Replace the defective IC if the sound returns. If not, the most critical voltage measurements are at pins 1, 2, 4, 10, and 11. Low-voltage measurements at pins 1 and 2 might indicate that an IC is leaky, as indicated by the small voltage in Fig. 11-13. Check the supply voltage at the 27-Ω supply resistor (26.1 V). A low supply voltage might indicate problems within the low-voltage power supply source or that a component is leaky.

A distorted sound IC stage might be caused by a defective speaker coupling capacitor, IC, or an improper alignment of L201. If, after checking the speaker coupling capacitor and making the voltage test, the distortion remains, take voltage measurements on the IC terminals. Touch up the sound coil (L201) if all voltage measurements are fairly normal. Select the proper alignment tool. If not, you might destroy the small metallic adjustment core. Turn the sound halfway up. Slightly rotate the core clockwise. Notice if the distortion clears. If not, rotate the core counter-clockwise a full turn. Slowly rotate the core until the sound is best on all stations. It is usually only a slight adjustment. Replace the IC if the voltages are normal and a quick sound adjustment does not clear the distortion. These sound tests can be made by the beginner or intermediate repairer.

Signal tracing for sound distortion can be made with the use of an external audio amp. These tests should be made by the intermediate or more advanced electronics repair person. Using the external audio amp, check the input sound signal at pin 15. This signal might be quite weak, so turn up the volume of the external amp. Now, check the sound at the output of IC201 (pin 2). Suspect that the IC is defective if the sound is normal at pin 15 and distorted at terminal 2. All TV sound circuits can be checked for weak or distorted sound with the use of an external audio amp.

Removing and replacing the IC

After determining that the IC is defective, install a new one. Locate pin 1 by the identification notch on one end of the IC. It is the first pin to the left of the notch when viewed from the top side of the IC (Fig. 11-14). Mark an X where pin 1 is located on the chassis. Some manufacturers have the IC terminal numbers printed on the bottom side of the PC board. Desolder all pin terminals at the bottom side of the PC board.

Select a piece of copper solder wick that is impregnated with soldering paste to help you remove the solder from around each terminal. This handy solder remover comes in a small roll, and is sold at many electronics stores. It is used to lift solder off transistors, capacitors, resistors, and IC terminals. Hold the solder wick flat along side a row of IC terminals. Likewise, go to the other side and remove the excess solder. Proceed by removing solder from both rows of terminals. Flick each terminal with the blade of a pocket knife to be sure that the connection is loose (Fig. 11-15).

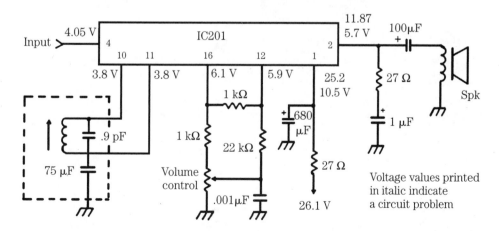

11-13 Check for an improper voltage on the IC terminals that indicates a defective IC. Low voltage on pins 1 and 2 indicates a leaky IC, as indicated with the voltages in italic.

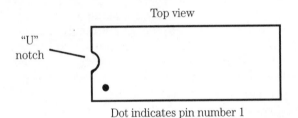

Top view

"U"
notch

Dot indicates pin number 1

11-14 Locate pin 1 by a dot, line, or "U" notch at one end.

Surface mounted
processors & IC's

11-15 Surface-mounted ICs or microprocessors mount flat on the PC wiring with gull-type wings.

Lift the defective IC with a pocket knife or the blade of a small screwdriver. Check where pin 1 goes and insert the new IC into the PC board holes. Be sure that all pins poke through each hole on the printed wiring side. Doublecheck each one. These small pins have a tendency to bend over and not fit into their respective holes. Solder each terminal with a low-wattage soldering iron or a battery-cordless soldering iron. Place the soldering-iron tip to one side of the wiring and pin connection. Apply solder on the opposite side.

Do not leave the iron in contact with the connection too long to prevent overheating the IC. Doublecheck the soldering of each connection. Sometimes a magnifying glass is handy to check for good soldered connections. The solder might adhere to one side of the pin and not cover the whole connection. Retouch the terminals that appear to have poorly soldered joints. Run the back edge of a pocket knife blade between each pin connection. Besides removing excess rosin, the blade will clean out any solder bridges between the IC terminals that could damage the new IC. If two connections are accidentally joined together, remove the excess solder with a piece of solder wick material. After replacing a few ICs, the beginner will soon acquire the experience to move into the intermediate class.

Isolation transformers

Before attempting to service any TV, plug the chassis in a variable or isolation transformer. The transformer provides isolation from the ac power line when servicing a hot TV (Fig. 11-16). Most TVs made today do not have a power transformer and operate directly from the power line, making the chassis hot. Although the power cord has a polarity plug, use the isolation transformer between the TV cord and power outlet. The isolation transformer eliminates shock hazards and prevents damage to power-line operated test equipment. Some of these variable (step-up/step-down) transformers have a tapped switch to vary the applied line voltage for the intermittent TV. A few of these transformers are fused, but many are not.

11-16 Plug the TV into the isolation transformer before connecting a substitute tuner or test instrument to the chassis.

Although your test equipment can be operated from an internal power transformer, just plug the TV into the isolation transformer. Choose an isolation transformer with variable taps or a continuous-voltage control (variac). The isolation transformer can very from 0 VA to 150 VA with one or more isolation outlets. Prevent TV damage of blowing fuses and silicon diodes, internal damage to the connected test instrument, and possible shock to the TV technician. Play it safe.

Trouble symptoms

This section is similar in content to Chapter 10. Some of the headings indicating TV symptoms were covered earlier. Nonetheless, this is not a repeat of any earlier material. This section deals with several common or typical symptoms that might cause problems in a TV. Scan through the list of troubles listed and locate the desired symptoms; then follow the recommended action (Fig. 11-17).

Dead set

First, check the circuit breaker or fuse for a no sound/no picture/no raster symptom. Replace a blown fuse with the exact current rating specified. The circuit breaker can be quickly checked by clipping a lead across its two terminals. Sometimes a fuse will blow or a circuit breaker open because of a brief overload in a transistor. Suspect a shorted or leaky silicon diode if a new fuse opens (Fig. 11-18).

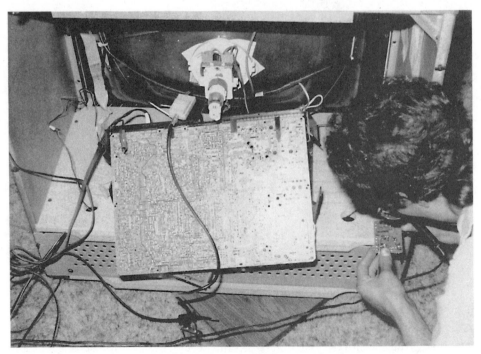

11-17 Isolate the symptom before testing or tearing into the chassis.

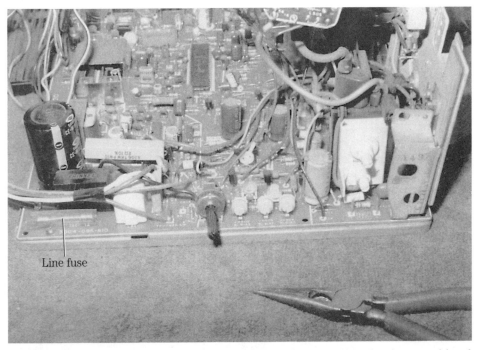

Line fuse

11-18 Locate the fuse near the rear of the TV chassis. Notice that this fuse is soldered to the PC board.

You might find four silicon diodes forming a bridge rectifier circuit, two individual diodes or one single diode in the power supply (Fig. 11-19). Each diode can be checked with the ohmmeter scale of a VOM or DMM. If you suspect any diode unsolder one of its leads and check it again. If lightning strikes a TV set, two or more diodes often become defective, along with an open fuse and voltage-isolation resistor.

If after replacing the open fuse, the set remains dead, check the isolation resistor (R701) for an open condition. Check each diode for leakage with the ac power cord removed. Then plug in the power cord and measure the dc voltage at the output terminals of the power supply. Suspect problems within the horizontal output circuit if the low-voltage power supply has a higher-than-normal low voltage.

Keeps blowing the fuse

Check each silicon diode if the fuse keeps blowing. The bridge rectifier is often four silicon diodes encapsulated into one component. Check each diode for leakage with the VOM or DMM. You can save fuses and possible damage to other components by inserting a 100-watt light bulb in series with the load. Just clip the lamp across the two fuse terminals, as illustrated in Fig. 11-20. The short is still present if the light has full brightness.

Often a leaky horizontal output transistor will cause most of the fuses in the TV to open. Remove the output transistor or disconnect its collector lead. Then plug in the power cord and observe the light. If the light goes out, replace the leaky

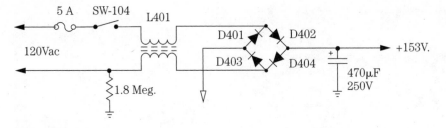

11-19 A bridge rectifier contains four separate silicon diodes inside one mounted component.

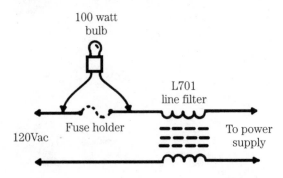

11-20 Clip a 100-watt bulb across the fuseholder terminal. It will remain bright with shorted diodes or a shorted filter capacitor.

horizontal output transistor. If the light remains bright, suspect that the flyback transformer is defective. Call in a professional electronic technician for a transformer replacement.

A shorted or leaky filter capacitor might cause the fuse to open. If the fuse blows with the horizontal output transistor removed, suspect problems within the low-voltage power supply. Either the voltage-regulation transistor is leaky or the filter capacitor is defective. Remove the positive lead of the first filter capacitor and run the lamp brightness test (Fig. 11-21).

To check for a blown ac fuse caused by a shorted automatic degaussing coil (ADG), remove the voltage isolation resistor or the input lead of the silicon rectifier. This will isolate the ADG circuit and prevent it from opening the fuse or keeping the 100-watt bulb bright. Automatic degaussing coils will sometimes short against the metal shield of the picture tube assembly. Simply unplug the ADG coil assembly if it is not wired directly into the circuit. The light will go out if the coil is not shorted.

Circuit breaker problems

If the circuit breaker keeps opening after resetting, suspect an overloaded condition or a defective circuit breaker. Clip another circuit breaker with the same amperage rating across the old breaker terminals. Reset the substitute circuit breaker several times. It will not hold if a true overloaded condition exists. Do not continue to push in the circuit breaker if the unit opens immediately after being reset.

Another method is to clip the 100-watt bulb test across the open circuit breaker. If the light remains bright, check for an overloaded condition in the power supply or horizontal output circuit. Check the low-voltage power supply for the probable cause if the circuit breaker is in the ac input circuit (Fig. 11-22). If the circuit breaker is in the horizontal output circuit, check for a leaky horizontal output transistor or damper diode.

Hum bars

Poor filtering of the low-voltage power supply can cause vertical roll of two or more black hum bars. An improper adjustment of the B+ control can also cause hum bars in the raster. Try to eliminate the black bars by adjusting the B+ control. Dried-up filter capacitors are a source of hum bars. To check, clip another equivalent value filter capacitor across the suspected one. Always, turn off the power and observe the correct capacitor polarity. Take the TV to a professional if these two tests do not remove the black bars from the raster.

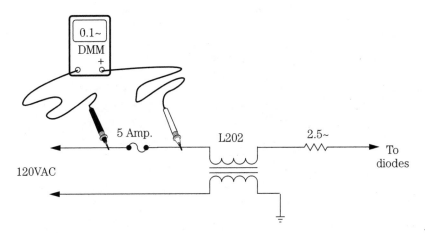

11-21 Check the fuse with a low-resistance continuity test and will be less than 1 OHM when normal.

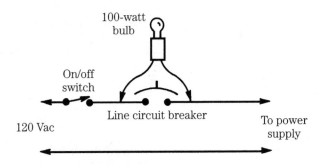

11-22 If the circuit breaker keeps popping open, clip a 100-watt lamp across its terminals.

Regulator hum bars

If hum bars are in the picture and cannot be eliminated with horizontal control or B+ control (at the rear of the chassis), suspect that trouble is in the low-voltage power supply regulator circuits. Take voltage measurements on the regulator transistor and compare the results with the schematic (Fig. 11-23). Often, a leaky regulator transistor will produce hum bars. Check for burned resistors around the regulator. Shut down the TV and shunt small electrolytic capacitors with the same or higher values.

No sound and no raster

A no-sound/no-raster problem might be caused by a defective low-voltage power supply or horizontal output circuit. The present-day TV will not have sound until the horizontal circuits are operating. Measure for the correct dc voltage between the collector terminal of the horizontal output transistor and the chassis ground. Higher-than-normal collector voltage can indicate that an HV regulator is defective or that a horizontal output transistor is open. High-voltage shutdown can occur with too high of a voltage at the collector terminal. Low collector voltage can be caused by a leaky output transistor or a defective low-voltage power supply.

A quick leakage test can be made between collector and chassis ground. A very low resistance might indicate that a transistor or damper diode is leaky. A low-resistance reading between the base and emitter terminals is normal for an in-circuit test. Remove the horizontal output transistor and test it out of the circuit. Check for a leaky damper diode if the horizontal output transistor are normal. Sometimes the output transistor is open between the base and emitter terminals. The transistor must be removed for this test. If in doubt, replace it with a new output transistor.

Chassis shut-down problems

Just about any component in the low-voltage power supply or horizontal circuit can cause chassis shutdown. Sometimes the chassis will not start up with a defective component. It's best to isolate the low-voltage power supply from the horizontal output

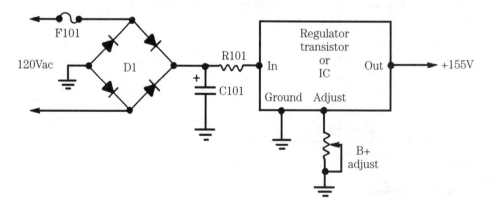

11-23 Hum in the picture can be caused by a leaky low-voltage regulator or filter capacitor.

circuits for chassis shutdown problems. Remove the B+ fuse going to the horizontal circuits. If a fuse is not used, remove the horizontal output transistor from the chassis.

Now, if the low voltage comes up to normal or goes higher, you can assume that the low-voltage circuits are normal. Check the horizontal circuits for defective components. The electronic technician checks the drive voltage with the scope at the collector of driver transistor and base of horizontal output transistor. Check the horizontal output transistor with a transistor tester or with the DMM diode tests.

If the chassis shuts down with a horizontal white line before shut down, suspect the vertical circuits. Most problems caused in the vertical circuits are caused by the vertical output transistors. Be careful when checking them in the circuit. Sometimes they will pop back to normal and test good. Simply remove the vertical transistor that receives the operating voltage from the low-voltage power supply. If the chassis comes up with a horizontal white line and does not shut down at once, check the vertical output circuits and transistors.

With IC vertical circuits, remove the V_{cc} operating voltage from that pin. Unsolder the V_{cc} pin or cut the foil of PC wiring feeding the supply voltage pin. Likewise, if the chassis does not shut down, suspect that the problem is in the vertical output IC and circuits. Do not forget to bridge cut PC wiring with bare hookup wire.

Checking horizontal circuits

Because the horizontal output transistor fails more than any other transistor in the TV, replacement of this transistor might solve the no-sound/no-raster problem. The transistor can be removed and a new one temporarily installed or the original can be tested out of the circuit. Before the body of the transistor can be removed, it is necessary to remove the two screws that secure the base to the chassis and to desolder the base and emitter terminals from the circuit. Some horizontal output transistors are plugged into a separate transistor socket and can be removed without desoldering (Fig. 11-24).

Check the transistor with a transistor tester, or take a resistance measurement with a VOM or a DMM. Low-resistance measurements in one direction are normal in a silicon horizontal output transistor (Fig. 11-25). A low-resistance reading in both directions between any two elements indicates that the transistor is leaky. A transistor can be open between any two elements, resulting in a no-raster/no-sound symptom with higher-than-normal voltage on the collector.

Replacing horizontal output transistors

If the horizontal output transistor is leaky or is suspected of breaking down with overload, replace it with a new one. Most power output transistors mounted on the metal chassis have a piece of insulation between transistor and chassis (Fig. 11-26). Remember, the horizontal transistor has the highest voltage supplied to it besides the picture-tube circuits. The metal body of the output transistor is the collector terminal with low voltage applied.

If the horizontal output transistor is mounted separately on a heatsink (Fig. 11-1), no insulator is found. The insulated PC board keeps the metal heatsink and transistor above ground. Sometimes the heatsink will come off when the two transistor mounting screws are removed.

11-24 A low-resistance measurement from the metal body to the chassis ground might indicate a transistor or damper diode is leaky.

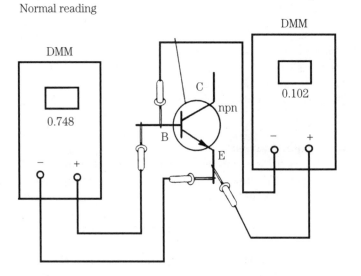

11-25 A low-resistance measurement in one direction can indicate that a horizontal output transistor is normal, but if it is low in both directions, the transistor is leaky.

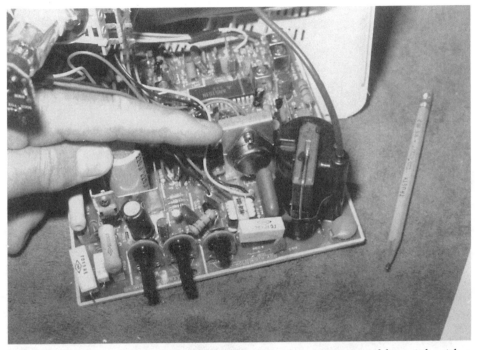

11-26 Be sure that the mica insulator is between the transistor and heatsink with a coating of silicon grease on both sides of insulator.

Place silicone grease on each side of the insulated piece and press it. Now, place the horizontal output transistor in the correct holes. You can reverse this transistor, except that the mounting holes will not match. Remove it, turn it 180 degrees, and insert it again. Be sure that the insulator ends are not broken off where the screws go into it because the transistor body could short against the metal chassis. Lightly tighten the two metal screws. Do not overtighten them because the transistor flange might "bite" into the chassis. Doublecheck with the low-resistance range of ohmmeter between the metal body of transistor and the metal chassis. When replacing horizontal output transistors mounted on separate heatsinks, always place a thin layer of silicone grease on the bottom of the transistor.

Horizontal lines
Horizontal lines can be the result of improper adjustment of the horizontal hold control, horizontal circuit drifting, or a defective horizontal oscillator circuit. Check the circuit and chassis layout for the existence of a horizontal hold control or oscillator coil. In many of the lower-priced models, the horizontal oscillator coil is also the horizontal hold control. Improper adjustment of this control will cause horizontal lines running diagonally to the left or right on the screen without a viewable picture (Fig. 11-27). Turn the control until the lines get wider apart. If the core of the coil begins to bind and cannot be turned any farther, remove the tab or stop around the plastic adjustment shaft. Now, rotate the core until the picture flops in. Don't go too far or the picture will form into horizontal lines of the opposite direction.

11-27 An improper adjustment of the horizontal hold or the horizontal oscillator is off frequency.

Suspect that a separate horizontal oscillator coil is the cause of drifting if a regular carbon control is used as a horizontal hold control. Locate the horizontal oscillator coil, which is usually located in a shielded container. The core could have a screwdriver adjustment or it might require a hexagon plastic tool to adjust it. If the core will not turn either way, heat the core by heating a metal hexagon tool that has been placed into the core. Use a soldering iron to heat the tool. Heat from a hair dryer or a regular heating tool can also help loosen the metal core area. In many cases, the core is locked in place with wax. Replacement of defective horizontal transistors and ICs should be left to the professional servicer.

Insufficient width

Look for a horizontal size or width control in the schematic TV parts layout. Adjust it for a normal-sized picture. Improper adjustment of the B+ control can cause the sides to pull in. Measure the B+ voltage at the collector terminal of the horizontal output transistor. A defective horizontal output transistor can cause insufficient width.

No raster, but sound OK

Suspect that the picture tube or circuits are defective if no raster or picture with normal sound and high voltage. If you have no high-voltage voltmeter, just listen for the yoke to expand when high voltage comes up. Hold the back of your hand near the screen of the picture tube and the hair should stand up, indicating that high voltage is present.

Check the heater or filaments of the CRT for the light of each gun at the rear of the tube. No light may indicate an open filament of CRT or no heater voltage. Shut down the chassis and remove the socket of the picture tube. Check the schematic for correct picture tube heater pins. Any resistance measurement between 1 and 50 Ω indicates that the filament or heater is normal. If there is no measurement, the heaters are open. A new CRT must be installed.

If the continuity of the picture tube filament is good, suspect that the socket is bad or that the connections are poorly soldered. Inspect the socket for burned or poorly soldered connections from heater pins and socket to the CRT board (Fig. 11-28). Take a continuity measurement across the two heater pins. No measurement indicates that the heater winding or wire connections are open. Closely inspect the wiring and soldered connections. Simply soldering the heater tube pin socket wires and chassis connecting wires can fix an open heater winding (Fig. 11-29).

Remember, the heater voltage is taken from a winding on the horizontal output transformer (flyback). Do not attempt to measure this ac voltage from the flyback transformer.

Beware of high voltage

Be extremely careful when working around the horizontal output transformer, high-voltage lead, or anode connections on the picture tube. You could receive a

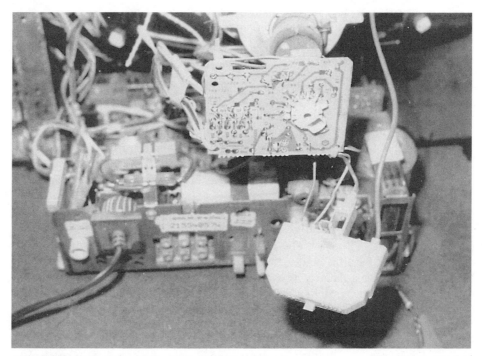

11-28 Check the CRT PC board for poorly soldered connections or burned heater pins if the filament of the picture tube will not light.

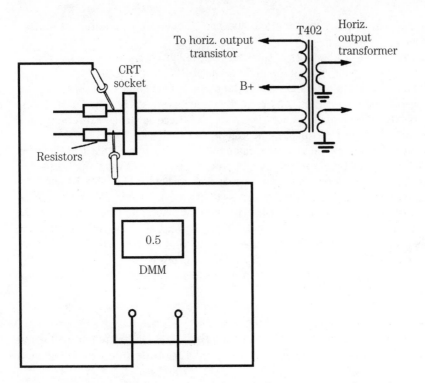

11-29 Be sure that the heater or filament winding is not open by sticking the ends of resistors into each heater pin of the CRT socket.

terrible burn or shock if you place your fingers or hand on the high-voltage connection (Fig. 11-30). The electronic technician always respects the high-voltage source of the TV when it is operating. If, for any reason, the chassis must be pulled out of the set for tests, the anode lead and socket must be discharged from the anode to the black aqueduct coating of the picture tube.

Be sure that the aqueduct coating of the CRT is properly grounded to the chassis. If not, arcing will develop and you can hear it in the TV's audio and see white dashes on the screen. You might receive a shock when touching the picture tube with an ungrounded tube.

No raster, but normal high voltage

A no-raster symptom with normal HV could result from video or picture tube problems. Check for a filament light in the rear end of the picture tube. No visible light might indicate a filament is open or that an improper voltage is at the heater terminals. Pull the ac cord. Remove the picture tube socket. Measure for continuity across the heater terminals. A no-resistance measurement indicates that the heater is open internally.

Replacing the picture tube is the only answer. Check the heater wiring from picture tube socket to the PC board for continuity. In older TVs, the picture tube heater received the voltage from a power transformer winding. Today, the heater voltage is derived from a voltage in the flyback circuitry. Use the low-resistance scale of a VOM to trace each heater wire back to its source. A broken PC wiring connection, tie-in wire, or an open isolation resistor might prevent the heater from lighting. Because video problems are very difficult to locate, leave them up to the professional TV technician.

Vertical circuits

An improper adjustment of the vertical height and linearity controls might cause the picture to roll. Adjust both controls with the vertical hold control to stop the picture from rolling. These controls are on the rear of newer black- and-white and older TVs. If the rolling does not lock in (on the newer type of TVs) without any vertical controls, check the vertical transistors and vertical sync circuits (Fig. 11-30). If the picture is rolling both horizontally and vertically, suspect a problem in the sync circuits. In the latest TVs, the sync circuits are usually contained in the deflection IC.

Only a white horizontal line

No vertical sweep might produce a single white horizontal line. Adjust the vertical hold, vertical size, and linearity controls to see if the sweep will return. You might find only one vertical adjustment control on your TV. Vertical circuits are very tricky to service. Replacing both vertical output transistors can solve many vertical sweep

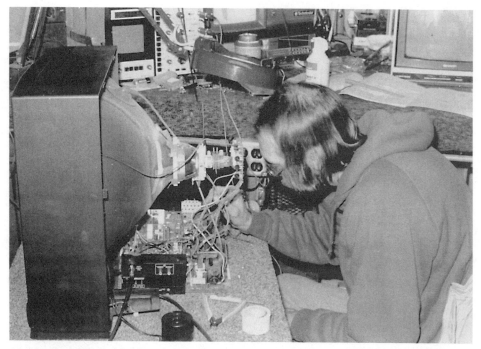

11-30 Be careful of high voltage when working around the picture tube anode lead and flyback.

problems. Measure the voltage on the output transistor. An improper voltage can be produced by an open resistor in the low-voltage power supply.

Vertical foldover

Excessive vertical foldover cannot be adjusted with the vertical height and linearity controls. Often, leaky or open vertical output transistors or a change in bias resistors might cause vertical foldover. Improper voltages on the bottom vertical transistor might cause foldover. Shunt each electrolytic capacitor in the vertical circuits with a similar value to see if foldover is caused by a dried-up capacitor. Sometimes, both of the vertical transistors must be replaced to cure excessive vertical foldover (Fig. 11-31).

No sound, but normal picture

Sound problems are fairly easy to service. The most troublesome components are the speaker, output transistors, ICs, speaker coupling capacitors, and bias resistors. A low hum in the speaker might indicate that the speaker is normal with a defect in the sound circuit. No sound or hum in the speaker can indicate that a speaker or coupling capacitor is open.

To check, touch the center terminal of the volume control with your finger and listen for hum in the speaker. Place a finger on the base terminal of the AF or driver amp in a transistorized sound circuit and check for a loud hum when the volume control is turned wide open. Clip a substitute speaker across the speaker

Separate heat sink with vertical output transistors

11-31 Suspect defective vertical transistors mounted on separate heatsinks if the height is improper, the screen keeps rolling, or a horizontal white line is present.

terminals. If you hear no sound from the speaker, take voltage and resistance measurement of the output transistor and IC.

Distorted sound

Leaky transistors, burned bias resistors, leaky ICs, damaged speakers, and improper sound alignment produce more distortion problems than all other components in the sound stages. Disconnect one speaker lead and temporarily clip another speaker into the circuit to verify whether the distortion is caused by a defective distorted speaker. Out-of-circuit transistor testing will provide more comprehensive results, although in-circuit testing can pinpoint any serious defects. Any marginal transistor operation could go undetected. Always check for a burned bias resistor after locating a leaky transistor.

A low and distorted sound stage can be signal traced with the aid of an external audio amplifier. This signal tracing method is especially helpful in checking input and output terminals of an IC sound circuit for distortion. Check for distortion at the input terminal of the IC. The signal at this point might be very weak. If the input signal is free of distortion, check the output terminal. Take voltage and resistance measurements on all of the IC terminals. Replace the IC if in doubt.

Chassis or HV shutdown

Component failure in the low-voltage power supply and horizontal circuits can cause HV or chassis shutdown. Shutdown circuits are circuits used in recent color TVs to prevent excessive radiation that otherwise could occur from excessive high voltage. Check the fuse and circuit breaker for the chassis shutdown circuit. An overloaded condition in the low-voltage power supply or horizontal output circuits can cause chassis shutdown (Fig. 11-32).

High-voltage shutdown can be caused by a higher-than-normal power line voltage, improper B+ from the power supply, or a malfunction of components in the high-voltage regulator and shutdown circuits. Locate the B+ or high-voltage control, which is located in the low-voltage power circuits (Fig. 11-33). Readjustment of the B+ can sometimes solve a high-voltage shutdown problem. Measure the voltage at the collector terminal of the horizontal output transistor and compare the results with the specified correct voltage. Reduce the voltage by adjusting the B+ control if the collector voltage measured slightly higher. Call in a professional technician if the measured voltage is very high compared to that listed on the schematic.

Only horizontal lines

If the horizontal lines cannot be eliminated with the horizontal oscillator control, suspect high-voltage shutdown. Measure the high voltage at the CRT anode socket with a high-voltage meter. Do not attempt to measure the picture-tube anode voltage with a regular VOM or DMM. This voltage could measure from 7.5 up to 31 kV. You would burn up the meter and receive a terrible shock from the high voltage.

In the older TVs, when higher voltages were placed on the picture tube because of the increase in size and greater brightness, the HV shutdown circuit was designed to indicate a higher voltage or shut down the TV (Fig. 11-34). The X-ray protection circuit is used in modern TVs to shutdown the chassis by disabling the

11-32 Check the different components on the TV that can overload and cause chassis shutdown.

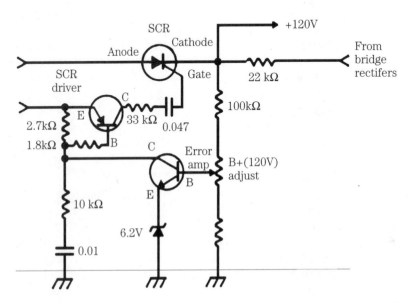

11-33 Improper high-voltage can result from an improper adjustment of the B+ HV control.

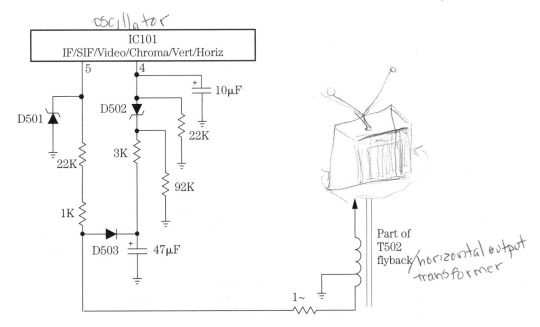

11-34 The x-ray or HV shutdown circuit takes a pulse from the flyback and feeds it to pins 4 and 5, causing the horizontal oscillator circuit inside of IC101 to malfunction.

horizontal oscillator or driver stage, if excessively high voltage is present. Remember, a defective X-ray protection circuit can shutdown the TV with normal high voltage at the picture tube.

Discharging the picture tube

The picture tube must be discharged before removing the chassis or when replacing a defective tube. For tube chassis, electronic technicians discharge the anode connection to chassis ground with screwdrivers. Most technicians use two long screwdrivers. Place one screwdriver blade against the aqueduct coating and slip the blade of the other under the rubber cover of the anode high-voltage cable (Fig. 11-35). Cross the screwdrivers so that they are touching when the blade reaches the anode button.

Do not try the screwdriver method on a solid-state chassis. You might accidentally destroy some transistors and ICs with this method. Always discharge the anode cable connection to the grounded aqueduct coating on the bell of the picture tube. Remove the high-voltage cable after discharging. Short the anode button to the aqueduct coating to completely discharge the high voltage.

Snowy picture

A poor antenna signal, a dirty or defective tuner, or defective AGC circuits all produce snowy pictures. The antenna signal can be checked with those of a second set. Notice if the stations come and go when the tuner knob is rotated. A dirty tuner can cause an intermittent or snowy color picture. First, clean the tuner contact points with a good cleaning spray or solvent. Most defective tuners are sent to a tuner repair depot for major cleaning and service.

Improper adjustment of the AGC control can cause a snowy picture. Some TVs have a regular AGC control and another delay AGC control. Watch the TV screen with a mirror as you adjust both controls to eliminate the snow from a local TV broadcast. Tune in a fringe station broadcast and verify your adjustments. If there is no obvious improvement in the picture quality after making the AGC adjustment, an AGC circuit, tuner, or IF stage might be defective (Fig. 11-36).

Dim and blotchy picture

If the picture remains dim after the screen brightness and subbright control are turned wide open, suspect that the picture tube is bad. A blotchy close-up of a person's face is a good indication that the picture tube is defective. This condition can occur with the brightness control turned fully up, then, when the contrast control is turned up, a patch of brightness appears.

The picture tube can be given a new lease on life with a filament booster or a rejuvenation charge. *Rejuvenation* is a cleaning process of the tube heater and cathode elements. Although these two methods of restoration of brightness very seldom last for more than six months to one year, life is added to the TV. The picture-tube booster or brightener plugs on the end of the picture tube, but the rejuvenation of a picture tube must be done at the radio-TV shop with an expensive test instrument.

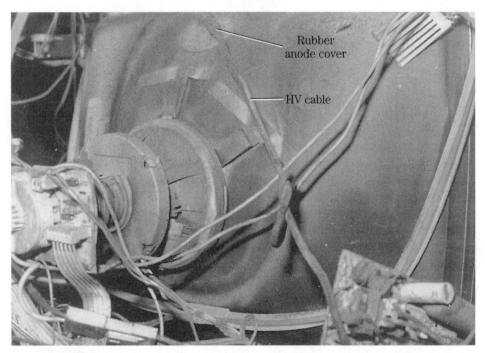

11-35 Discharge the anode button with two long metal insulated screwdrivers to the adquadag at the outside of the picture tube.

11-36 Snow can be caused by an open transmission line, or a defective antenna tuner, IF, or AGC circuits.

Poor color

Most color circuit problems must be serviced by a radio-TV technician. You can save a service call or a trip to the shop by rechecking the fine-tuning adjustment of the tuner. The fine-tuning control is probably located right behind the tuner knob. Be sure that the color control is turned up with the tint control set in the middle of rotation. Intermittent color can be caused by a dirty tuner. Readjust the color-killer control if one is on the back of the chassis.

Color spots in the picture

If the carpet sweeper is run close to the TV and shut off, the screen might become magnetized, and display unwanted color spots. Placing large stereo speakers near the TV cabinet can do the same thing. In most TVs, each time the chassis is turned on, an automatic degaussing circuit functions for a few seconds. The screen is demagnetized once again, removing any patches of coloration from the raster.

Suspect a defective ADG circuit if you notice stray spots of color in several areas of the screen. The raster at the corners and the bottom half of the picture might have a discoloration. Check for an open degaussing coil or plug. An open or broken thermistor can cause the degaussing circuit not to function. The degaussing circuit is very simple and can be checked with the low-resistance scale of the VOM (Fig. 11-37).

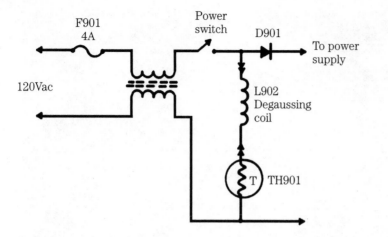

11-37 Check the degaussing circuit with the low-resistance scale of the VOM or DMM.

Arcing

High-voltage arcover is the most common of the arcing noises heard in a TV. The arcing can occur at the anode connection of the picture tube, high-voltage flyback, transformer, high-voltage socket, and tripler unit. Be very careful when high voltages are arcing inside of the TV. With the back cover removed, you can quickly see where the arcing is occurring.

Before working around the high voltage, discharge the picture tube. Take a long and short screwdriver, touch one of the screwdriver blades to the metal chassis or shield of the picture tube and slip the blade of the other screwdriver under the rubber cover and touch the high-voltage anode connection of the picture tube. Touch both screwdriver blades together. Keep your hands on the insulated handles of the screwdriver. Hold the discharge for 30 seconds.

Cleaning off the high-voltage connection at the picture tube anode might solve arcing at the socket. Wipe off the anode connection with a clean cloth and isopropyl alcohol. Let the surface dry completely before replacing the anode socket. Be sure that both wire clips are free from rust and are securely caught inside the socket. A poor connection here might make the arcing worse. Arcing indications in the raster can be caused by a poor ground between the picture tube and the TV chassis. Clip an alligator lead from the metal chassis to picture tube shield or mounting bolts and notice if the arcing indications disappear. If so, install a shorting wire from the mounting bolt of the picture tube to a metal screw on the metal chassis.

Tripler arcover

The high-voltage tripler unit might arc over between the tripler and chassis ground. Often, diodes within the tripler will short or high-voltage capacitors open and cause arcover. You can hear a loud arcing sound if the tripler is defective. Sometimes the defective tripler will arc internally and overload the flyback transformer. Shut the set off and feel the body of the tripler. These triplers should run cool, but if it is warm or hot, replace it. Replace the tripler unit if you notice arcing around or beneath the unit. Tripler units must be replaced with exact replacements. Make note of where each lead goes before removing the old tripler.

You cannot repair these units with high-voltage arcover spray or putty. You are just wasting time and energy. Using these techniques, the arcover mighty quit, but it will resume before long.

What not to touch

There are many circuits within the TV chassis that the electronic beginner should not touch without the correct knowledge and test equipment. You can only make more problems and damage the TV. Here are the primary circuits or procedures where beginning electronic technicians should use caution:

- Critical horizontal components
- High-voltage measurements
- High-voltage circuits
- High-voltage regulator circuits
- Replacement of the picture tube
- Critical chopper and switched-mode power circuits
- System control circuits
- On-screen displays
- TV receiver alignment
- Do not attempt to turn screws or coil cores to improve the picture without the proper test equipment
- Removing and replacing large ICs
- Removing and replacing microcomputer chips
- Video and sync circuits
- Stereo audio circuits

On the other hand, if you gain knowledge from reading this book or obtain on-the-job training, go as far as you can to put that TV back into operation. You can do many of the repairs in this book, but be careful not to damage any components or receive a shock when taking critical voltage measurements and working around the picture tube with it operating.

Lightning damage

Lightning can damage the antenna lead-in, power transformer, or power-supply circuit. First, check the fuse and silicon diodes for possible damage (Fig. 11-38). Look for a large voltage-dropping resistor that has become open. Inspect the ac cord and plug for possible damage. Check for burned PC wiring in the power supply circuit.

Most newer TVs have a capacitor in each antenna terminal to protect the balun coils (Fig. 11-39). In older TVs, the balun coils are wired right into the antenna coil circuit. Inspect the antenna lead-in and surrounding circuits for burned-off terminal wires and components. After lightning damage has occurred, suspect a defective tuner if the picture is snowy.

Lightning can strike the antenna and burn out sections of the lead-in wire. Cable TV can also be damaged by lightning. Inspect the cable where it connects to the antenna transmission line. If there are signs of lightning damage, replace the entire

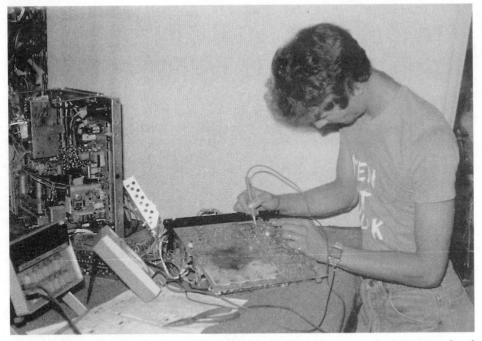

11-38 Lightning could destroy the TV. Notice the black area on the bottom side of the chassis.

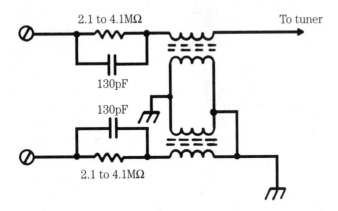

11-39 A small capacitor placed in each antenna leg can prevent lightning from entering the balun coils or tuner.

transmission line, whether it is 300-Ω twin lead or shielded 72-Ω coaxial cable. Check all matching transformers and boosters connected to the transmission line.

Remove the ac plug and disconnect the antenna cable when you go on vacation or when the TV will not be in use for several days at a time. Many of the new chassis have a surge voltage protector wired across the ac line. These line-surge varistors can be wired directly into the power-line circuit or a separate protector unit can plug directly into the ac power outlet. Then, the TV's line cord is plugged into the protection device.

12
CHAPTER

Symptoms and causes

This chapter lists several common trouble symptoms and things that might cause such a malfunction. Remember that these are typical. They are intended to guide you in the right direction so that you can possibly pinpoint the exact trouble after checking over your TV.

You have seen some of the headings indicating TV symptoms earlier in this book. Nonetheless, they are presented here as an aid to refresh your memory after once reading this book. Just thumb through these pages to locate the trouble. Then read through the listed possible causes, one by one, testing your receiver for defects.

When troubleshooting a TV, always keep a mental image of the block diagram in your mind. Better yet, keep the actual diagram at your side. You'll find that by referring to the block diagram, you can isolate the defective area much easier.

Likewise, when removing several components to get at the defective one, mark down how they were removed in succession and reverse the procedure. When removing and replacing IC, mark on the PC board where terminal one is located. On a piece of paper, mark all terminals of the suspected transistor before removing it from the board. The same applies to diodes and electrolytic capacitors, to watch for correct polarity. Replace damper and HV diodes with components that have the exact voltage ratings.

Always replace capacitors with those that have the same working voltage or higher. Electrolytic capacitors can be replaced with parts that have a higher capacitance and voltage if the exact value is not available. If one filter capacitor is leaky, but others in the same can test OK, replace the whole can. If one goes, they all will go within a few months. Do not solder capacitors into the circuit with long leads because the part will flop around and get in the way.

Do not replace a $\frac{1}{2}$-watt resistor with a $\frac{1}{4}$-watt one. It's best to use a 1-watt size, if available. Try to replace all fuses with the same amperage rating. Do not wire across the fuse without providing protection to the TV chassis. Use common sense and servicing precautions while working on TVs.

Set completely dead (Fig. 12-1)

- Check ac outlet for power with a table lamp
- Check the line cord
- Check the line cord interlock at rear of the TV
- Check the line or main fuse in the TV
- Check the circuit breaker
- Check the B+ fuse
- See if the dial lights
- Check the on/off switch with a DMM
- Check the silicon diodes or bridge rectifiers
- Measure the dc voltage at the main filter capacitor
- Notice if the picture tube lights

Keeps blowing the line fuse (Fig. 12-2)

- Place a 100-watt bulb across the open fuse for an indicator
- Check second the B+ fuse
- Remove the horizontal output transistor to see if the horizontal circuits are blowing the fuse
- Check for a correct dc voltage at the main filter capacitor
- Check the diodes and the filter capacitor if the B+ fuse does not open
- Check for shorted a ADG coil against picture tube metal frame
- Check for a leaky damper or horizontal output transistor if the low-voltage source is normal

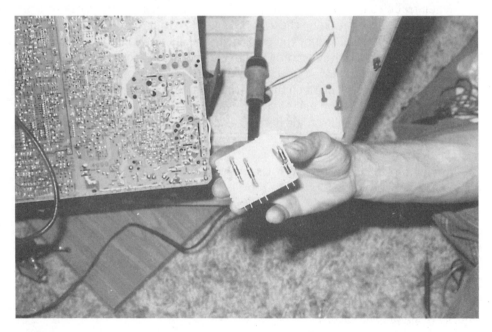

12-1 Check the silicon diodes in the power supply of an RCA CTC140 chassis.

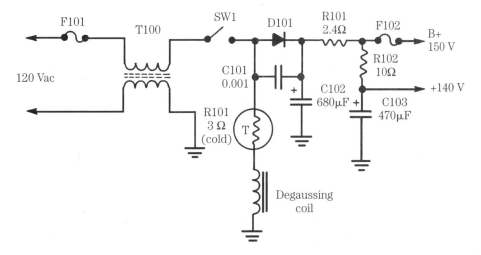

12-2 Check F101, D101, C102, and the degaussing coil if the line fuse (F101) keeps blowing.

No sound/no video/raster OK

- Check the tuner voltage
- Check the shielded cable from the tuner to the main chassis
- Clean the tuner with a spray lubricant
- Check the AGC adjustment
- Check the transmission line from the back of the receiver to the tuner
- If the picture is snowy, check for a broken antenna lead
- If one or two channels are not visible, suspect that the tuner is dirty

Snowy picture

- Check the outside antenna and transmission line
- Check the antenna connections on the TV input terminals
- Clean the tuner with tuner spray
- Check for burned or open balun coils
- Readjust the AGC
- Push the IF cable into the socket
- Substitute a tuner and notice if the varactor tuner is defective
- Notice if the antenna has been turned by the wind (wrong direction)

No sound/normal picture

- Substitute another speaker
- Check the speaker-wire connections
- Check the hum at the center tap on the volume control to see if the sound problems are in the output circuits

- Check the sound discriminator coil for correct sound
- Test the audio output transistors
- Take critical voltage and resistance measurements at the sound IC
- Replace the defective output transistor or IC with a component that has the part number, if possible

Weak sound (Fig. 12-3)

- Substitute another speaker
- Shunt another electrolytic capacitor across the capacitor tied to the speaker
- Test and replace the output transistors
- Replace any defective ICs

Noisy sound

- Determine if the noise is in the output or input audio stages (turn down the volume control).
- If noise is still heard, check the output transistors and the IC
- Replace the output IC if it has a low frying noise
- If the sound is distorted, check the speaker and output circuits

Distorted sound

- Check the speakers
- Readjust the discriminator coil
- Replace the output transistor
- Replace the output IC

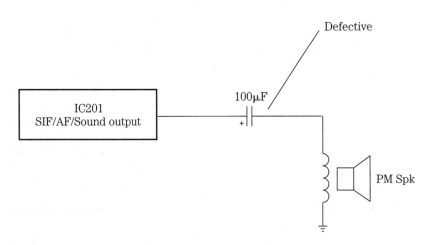

12-3 Clip another electrolytic across the coupling capacitor if the audio is weak or intermittent.

Intermittent sound

- Check the speaker
- Check the speaker-coupling capacitor
- Replace the intermittent transistor
- Replace the intermittent IC
- Spray cleaning liquid on the volume control
- Check the speaker and volume control leads.

Sound OK/no raster (Fig. 12-4)

- Notice if the filaments light in the picture tube
- Check for an open B+ fuse or resistor
- Check if the high voltage comes up—the hair on your arm will stand up near the TV picture tube
- Check the horizontal hold-control adjustment
- Check the dc voltage at the main filter capacitor
- Check from the horizontal output transistor (body) to the chassis ground for leakage
- Check and test the horizontal output transistor out of the circuit for leakage
- Measure the dc voltage-to-collector terminal of the horizontal output transistor while it is out of the circuit
- Check for a leaky damper diode with the diode test of the DMM

Hum bars in raster or picture (Fig. 12-5)

- Notice if hum is also in the speaker
- Discharge and shunt the main filter capacitor with a good electrolytic capacitor that has the same capacity and working voltage
- Notice if the hum disappears with the shunted capacitor
- Readjust the B+ or HV control
- Check the regulator transistor and diode for leakage

Loud arcing or popping noise (Fig. 12-6)

- Check the high-voltage lead to picture tube connection
- Check if the anode rubber-cup connection is defective
- Check the high-voltage lead for breaks
- Check the neck of the picture tube (blue arc) for glass cracked inside
- Inspect the ground strap on the CRT
- Check the tripler for arcing-through to the metal of the chassis
- If the speaker pops, check for a defective audio output transistor
- Inspect the IHVT flyback for HV diodes arcing over in the molded case

Poor width

- Adjust the B+ control
- Check the low-voltage source to the horizontal circuits
- Test the horizontal output transistor

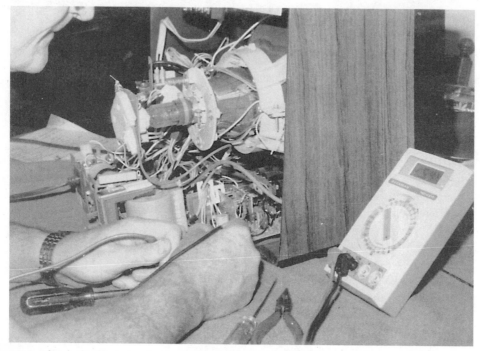

12-4 Check the horizontal circuits if the B+ or secondary fuse opens.

12-5 Hum bars in the raster can be caused by the main filter capacitor or by a leaky video transistor.

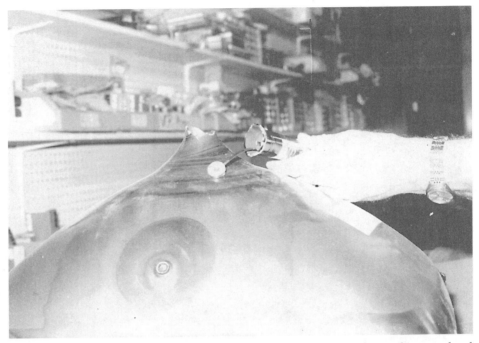

12-6 Sometimes the picture tube will crack inside the yoke and cause loud arcing noises.

- Check the horizontal drive waveform on the horizontal driver transistor
- Replace the damper diode
- Check for a defective pin-cushion component
- Check for a leaky regulator transistor or diode

No horizontal or vertical sweep (Fig. 12-7)

- Check the B+ voltage to the vertical and horizontal circuits
- Disconnect the yoke lead plug if a white area is in the middle of the screen
- Check the low-voltage power supply
- Round circle-check position of the yoke on the neck of the picture tube

Horizontal lines in the picture

- The horizontal oscillator is off frequency
- Adjust the horizontal hold control, if one is available
- In older chassis, the HV is too high, so disable the horizontal circuits
- The horizontal transistor is defective
- The horizontal deflection IC is defective
- Change the value of capacitors in the horizontal oscillator circuits

Horizontal foldover (Fig. 12-8)

- Replace the horizontal output transistor
- Check the horizontal deflection yoke

12-7 A white dot or line in the center indicates an that the yoke socket or plug is open.

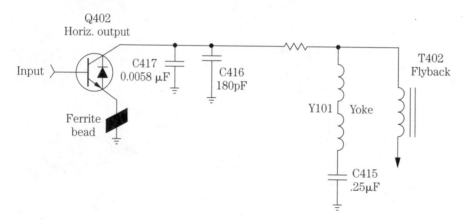

12-8 Check the following parts for horizontal foldover.

- Check capacitors C417, C416, and C415 (in this example)
- Check for a defective part in the pin-cushion circuits

Chassis shutdown

- Check for a shorted picture-tube gun assembly
- Check for a defective horizontal output transistor
- Check for a poor connection on the horizontal driver transistor
- Check for a defective driver transformer
- Check for a leaky component in low-voltage power supply

- Check for a leaky capacitor
- Check for excessive high voltage
- Check for an excessive voltage source to the H.O.T.
- Check for a defective regulator transistor or diode
- Check for an excessive power line voltage (typically, 117 to 120 Vac in North America)

Horizontal white line (Fig. 12-9)

- Defective vertical sweep circuits
- Open vertical yoke winding
- Open ground-return resistor
- Check the vertical size and linearity controls
- Check the vertical leads to the yoke assembly
- Test the vertical output transistors
- Take critical voltage tests at the vertical-output IC
- The vertical yoke coupling capacitor is open
- Remove the vertical output transistors for accurate testing
- Inspect the bias resistors while the vertical output transistors are out of the circuit

Improper vertical sweep (Fig. 12-10)

- Check the low-voltage source to the vertical circuits
- The vertical output transistor is leaky

12-9 A horizontal white line indicates that there is no vertical sweep.

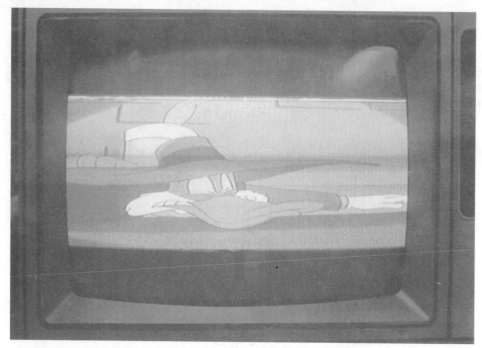

12-10 Improper vertical sweep can result from leaky output transistors or ICs, burned bias resistors, or improper voltage source.

- Check for a leaky vertical-output IC
- Check for burned bias resistors
- Check for a leaky deflection yoke
- Check for a leaky vertical coupling capacitor to the yoke assembly
- Check the vertical size and height control
- Check the vertical centering control, if available
- Check for burned spots in the vertical controls
- Shunt large capacitors in the vertical circuits
- Check for leaky capacitors on the vertical-output IC terminals
- If only a keystone-shaped, slanted raster is showing, the yoke is defective

Vertically rolling (Fig. 12-11)

- Check the vertical-sync circuits
- Check the vertical-hold circuits
- Check the adjustments for vertical height and linearity
- Check for excessive noise on the low TV bands
- Check for poor antenna or transmission line
- Check the AGC control setting
- Clean the tuner with cleaning spray
- Check that the outside components are tied to the vertical circuits
- Notice if the picture locks in horizontally

- Check the vertical voltage source
- Shunt the main filter capacitor with vertical crawling

Black top half or bottom

- Check the vertical-size and linearity adjustments
- Check the top vertical output transistor for black top
- Check the bottom vertical output transistor for black bottom
- Replace both vertical output transistors if one is leaky
- Replace the vertical output IC
- Check the vertical driver transistor if heavy retrace lines are at the top
- Replace any burned vertical bias resistors

Bunching vertical lines (Fig. 12-12)

- Replace the vertical output transistors
- Replace the vertical output IC
- Replace the vertical burned bias resistors
- Replace the leaky vertical diodes
- Check the resistance of all high-resistance resistors in the vertical circuits
- Shunt the vertical yoke electrolytic coupling capacitor
- Remove one end of each bias resistor and check for open resistance

12-11 Check for poor vertical sync, vertical hold circuits, and improper voltage source to the vertical circuits if there is vertical rolling.

12-12 Poor vertical linearity and bunching of lines at the top can be caused by defective vertical output transistors, ICs, resistors, and electrolytic capacitors.

Vertical foldover

- Replace defective vertical output transistors
- Replace the defective vertical output IC
- Check the resistance of the bias resistors
- Check all electrolytic capacitors in the vertical feedback circuits
- Check the yoke-return resistor
- Check the yoke-return electrolytic capacitor
- Test each bias resistor for correct value
- Replace both vertical output transistors if one is leaky
- Check for leaky capacitors and resistors in the pin-cushion circuits
- Resolder all pin-cushion coil and transformer connections

No color

- Check the color oscillator transistor
- Be sure that the black-and-white picture is very good
- Check for a defective 3.58-MHz crystal
- Check for a defective color IC
- Scope the color circuits
- Check for a defective color input capacitor to the color IC
- Check for a misadjustment of the color control
- Check for an improper low-voltage source to the color circuits

- Check for an improper voltage to the IC supply pin (V_{cc})
- Check for a leaky bypass capacitor on the color IC
- Adjust the color killer, if available
- If one or two channels does not have color, suspect that the tuner is dirty

Weak color (Fig. 12-13)

- Check for a good black-and-white picture
- Check the low-voltage source to color circuits
- Check for a leaky color IC or transistor
- Check for leaky or open diodes in the color circuits
- Check for burned resistors in the color circuits
- Check for open or leaky capacitors tied to the color IC
- Check for open coils or choke coils in the color circuits
- Do not overlook poor wiring board connections

Intermittent color (Fig. 12-14)

- Inspect for poor board connections at the color circuits
- Solder all connections in the color circuits
- Solder all pins on the color IC
- Spray each color transistor, capacitor, and resistor with coolant
- Inspect for poor eyelet or griplet connections

12-13 Take voltage measurements on the color IC terminal connections if the picture color is intermittent.

12-14 Solder all board, transistor, and IC terminal connections if the picture color is intermittent.

- Check the color-gun assembly of the picture tube (one color)
- Check for a defective 3.58-MHz crystal
- AFPC alignment (professional)

Barber pole or color sync

- Check the 3.58-MHz crystal
- Check the color killer
- AFPC alignment (professional)

Improper tint-control range

- Check the leads on the tint control
- Check for color-alignment problems
- Check for a defective color IC
- Check for improper voltage at the tint control
- Check all bypass capacitors at the color IC

No red

- Check the setting of the red screen control
- Check the color demodulator signal from the R-Y IC pin terminal
- Check the dc voltage on the red output color amp transistor
- Test for a leaky or open red output transistor

- Check the color bias and output transistor
- Test the picture tube
- Readjust the red screen control

No green or blue

- See the previous list, No red

All-red screen

- Check for a defective red gun assembly in the CRT
- Check for a leaky red bias transistor
- Check for a leaky red output transistor
- Check for an improper voltage on the red output transistor
- Readjust all screens and bias controls

Horizontal weave with color bars (Fig. 12-15)

- Check for a defective degaussing circuit
- Check for a defective low-voltage power supply
- Replace thermistor, if the degaussing circuit is on all of the time
- Check for an open resistor in the degaussing circuits
- Shunt filter capacitors if there is poor filtering in the power supply
- Readjust the B+ control
- Check for a defective low-voltage regulator transistor, zener diode, or IC

Smoking chassis (Fig. 12-16)

- Pull the ac cord and try to locate the overheated part
- Check for a smoking flyback

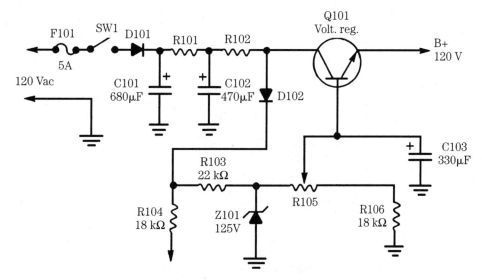

12-15 Wavy bars in the picture can be caused by improper B+ adjustments, or leaky voltage regulators or diodes in the low-voltage power supply.

12-16 Smoke curling up from the yoke assembly indicates that the coil windings are arcing internally. Remove and replace the yoke.

- Check for a smoking yoke assembly
- Inspect the power transformer
- Check for overheated large-wattage resistors
- Check for burned resistors
- Check for liquid spilled inside the chassis
- Check for arcing on the PC board
- Replace burned screen or focus controls
- Replace the arcover tripler unit
- Check for arcing under the CRT rubber cup

Poor focus (Fig. 12-17)

- Check the focus voltage on the focus pin of the CRT
- Check the screen/focus-resistor network
- Readjust the focus control
- Replace the defective focus control
- Check for a corroded focus pin at the CRT socket
- Clean the focus pin on the picture tube
- Check for a clogged focus spark gap

Sound beat pattern in the picture

- Check for an improper fine-tuning adjustment
- Check for an improper operation of cable TV

- Check for an alignment problem (professional)
- Check for a defective 4.5-MHz sound trap
- Readjust the sound trap

Ac hum bar floating in the picture

- Check the low-voltage power supply
- Check for a picture tube problem
- See if the antenna lines run too close to the ac power lines

Hum bars

- Check for defective degaussing circuits
- Check the defective regulator circuit
- Check for a leaky regulator transistor
- Readjust the B+ control
- Shunt the filter capacitor in the low-voltage power supply
- Check for a leaky horizontal driver and output transistor
- Check for a high-voltage power line near the outside antenna

No remote control (Fig. 12-18)

- Check the batteries in the remote control
- Substitute another remote to see if the remote is defective

12-17 Check the CRT socket for adequate focus voltage and to see if the focus pin is corroded.

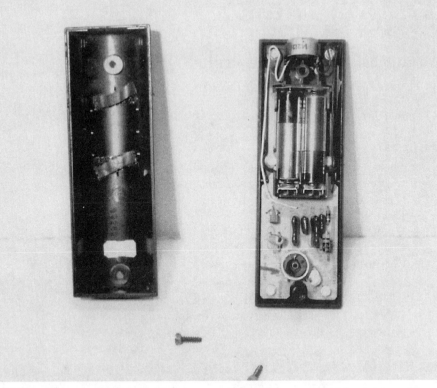

12-18 Check the batteries in the remote transmitter if one or more functions will not operate.

- Check for broken parts if the remote has been dropped
- Wipe off the ends of the batteries with a cloth
- Some remotes will not operate if the batteries are less than 0.5 volts
- Replace all batteries if one or more is weak
- Test the remote against an infrared indicator card
- Test the remote against portable radio (noise is generated in the radio, if the remote is working)
- Take the remote to a professional for service

Very little brightness

- Test for a weak picture tube
- Check the high voltage
- Check the focus voltage
- Check the voltage source to and at the video transistors and IC
- Check for a filament light in the CRT
- Check the luminance circuits (professional)
- Check for an improper screen voltage at the CRT

Retrace lines

- Check for a shorted or leaky picture tube
- Check for an improper adjustment of the screens and bias controls
- Check for adequate screen and boost voltages at the CRT
- Check for a correct low-voltage source at the color output transistors
- Take the luminance circuits to a professional to have them checked
- Check the vertical circuits

TV ghost in the picture

- Check the fine tuning
- Be sure that the ghost picture exists on all channels
- Check for a defective cable system (call the cable company)
- Check for a defective booster system
- Check for a defective tuner
- Check for poor IF cable connections
- Check the direction of outside antenna
- Check for an open in the transmission line
- Check for loose connections on the shielded cable
- Try another portable and notice if the ghost image appears on both TVs

Glossary

ac Alternating current. The type of electricity normally used in homes and most industries. Its contrasting opposite is direct current (dc), now obsolete except for certain specialized applications. All batteries supply dc.

ACC Automatic color control. A circuit similar in function and purpose to AGC, except that it is supplied exclusively to the color bandpass amplifiers to maintain constant signals.

ac hum A low-pitch sound heard whenever ac power is converted into sound, intentionally or accidentally. The common ac hum is a 60-hertz note.

AFC Automatic frequency control. A method of maintaining the frequency or timing of an electrical signal in precise agreement with some standard. In FM receivers, AFC keeps the receiver tuned exactly to the desired station. In TV, horizontal AFC keeps the individual elements or particles of the picture information in precise register with the picture transmitted by the TV station.

AGC Automatic gain control. A system that automatically holds the level or strength of a signal (picture or sound) at a predetermined level, compensating for variations caused by fading, etc.

amplifier As applied to electronics, a magnifier. A simple tube or transistor or a complete assembly of tubes or transistors and other components can function as an amplifier of either electric voltage or current.

antenna A self-contained dipole or outside device to collect the broadcast signal from the TV station. The collected signal is fed to the TV set with a shielded or unshielded lead-in wire.

anode The positive (+) element of a two-element device such as a vacuum tube or a semiconductor diode. In a television tube, an anode is an element having a relatively high positive voltage applied to it.

aperture mask An opaque disk behind the faceplate of a color picture tube; it has a precise pattern of holes through which the electron beams are directed to the color dots on the screen.

arc An electric spark that jumps (usually caused by a defect) between two points in a circuit which are supposed to be insulated from each other, but not adequately so.

aspect ratio   The relation or proportion between the width and height of a transmitted TV scene. The standard aspect ratio is 4:3, meaning that the picture is 3 inches high for every 4 inches of width, or four-thirds as wide as it is high.

audio   Any sound (mechanical) or sound frequency (electrical) that is capable of being heard is considered as audio. Generally, this includes frequencies between about 20 and 20,000 hertz.

B+   Supply voltage, as low as 1 volt dc in transistor circuits and as high as hundreds of volts in tube circuits, which is essential to normal operation of these devices. The plus sign indicates the polarity.

B+ boost   A circuit in TVs that adds to, or boosts, the basic B+ voltage. The boost source is a by-product of the horizontal deflection system. Also see *damper*.

balun coil   A set of balun coils are used between the antenna connection and the tuner to match the input of 300 Ω to 75 Ω at the tuner input.

bandpass amplifier   In a color TV, one or two color signal amplifiers located at the beginning of the color portion; they are designed to amplify only the required color frequencies. They pass a certain band of frequencies.

Barkhausen   A term applied to a display of one or two black vertical lines on the left side of the picture tube caused by some spurious behavior (oscillation) in the circuit. These lines are usually seen best when there is no picture on the screen (just a blank raster).

blanking   A term used to describe the process that prevents certain lines and symbols, which are required for keeping the picture in step with the transmitter, from being seen on the TV screen.

bridge rectifier   Four diodes are wired in a series circuit to provide full-wave rectification of a two-lead power transformer. The ac/dc TV chassis mighty use a bridge rectifier after the line fuse.

brightness   Refers to both the amount of illumination on the screen (other than picture strength) and the control that is used to adjust the brightness level.

burst   In color TV, a precise timing signal. It is not continuous, but comes in spaced bursts. It is transmitted to control the 3.58-MHz oscillator, which is essential for color reception.

burst oscillator   The precision 3.58-MHz oscillator, which is vital to color reception. It is kept in step (sync) by the burst.

buzz   This is sometimes called *intercarrier buzz*, a raspy version of ac hum, usually caused by improper adjustment of some IF circuits.

B-Y   The blue component of a color picture minus the monochrome.

capacitance   A measure of a capacitor's ability to store electrical energy. The capacitor was called a *condenser* at one time. Bypass and electrolytic filter capacitors are used in many TV circuits. The unit of capacitance is the farad.

carrier   The radio signal that carries the sound or picture information from the transmitter to the receiver. The carrier frequency is the identifying frequency of the station (e.g., 880 kHz, 93.1 MHz, etc.)

cathode   The negative or minus element of a two-element tube or semiconductor. The cathode and the anode combine to form a diode (two-element device). The cathode is also the source of electrons in such devices.

cathode-ray tube   A tube in which electrical energy is converted to light. An electron beam (or beams), originating at the cathode, impinges upon a phosphor light-emitting screen. TV picture tubes, radar tubes, tuning eyes in some FM sets, and many similar types are basically cathode-ray tubes.

chassis   The base where the majority of electronic components are mounted. The metal chassis might be common ground. Today, in the solid-state chassis, the PC board wiring is the main chassis.

cheater cord   An ac line cord for operating the TV without the back cover or the cabinet when troubleshooting and repairing. The original cord is attached to the back of the cabinet as a safety measure.

chroma   Another term for color. Color amplifiers are often called *chroma amplifiers*. The term is also used to denote the control used to increase or reduce the color content of a picture.

chopper circuit   The chopper power supply is a pulse-width-modulated (PWM), regulated power supply. The chopper supply circuits are quite similar to the horizontal deflection system.

circuit breaker   The circuit breaker might be used in place of the line fuse, which opens when an overload is in the TV circuits. Some horizontal output tubes have a separate circuit breaker in the cathode circuit.

clipper   A term that describes the operation of one of the sync circuits in a TV. This stage (tube or transistor) separates the sync (timing) signals from the picture information.

color bar generator   The color bar generator provides patterns for color alignment and color TV adjustments. Some of the NTSC generators have from 8 to 10 different patterns.

color killer   A special circuit whose function is to turn off the color amplifier circuits when a black-and-white signal is being received. Also, it is the control used to adjust the operation of the circuit.

comb filter   The comb filter circuit separates the luminance (brightness) and chroma (color) video information, eliminating cross-color that can occur in other chassis.

contrast   The depth of difference between light and dark portions of a TV scene. Also, the name given to the control for adjusting the contrast level.

convergence   The system that brings the three electron beams together in a color picture tube so that they all pass through the same hole in the shadow mask and strike the correct dots on the screen.

converter   A stage in the tuner or front end of a TV or any radio receiver that converts an incoming signal to a predetermined frequency, called the *intermediate frequency (IF)*. All incoming signals are converted to the same IF.

corona   Similar to an electric arc, except that this is a characteristic of much higher voltages (thousands). Corona occurs as a continuous, fine electrical path through air between two points, sometimes accompanied by a faint violet glow, usually near the picture tube.

crystal   A quartz of synthetic mineral-like slab or wafer that can vibrate at a precise rate or frequency. Each crystal is cut to vibrate at the desired frequency. Such a crystal is used in the 3.58-MHz oscillator to control its frequency.

CRT Cathode-ray tube; another name for the color picture tube.

damper A diode, tube, or semiconductor used in horizontal amplifier circuits to suppress certain electrical activity. It incidentally provides B+ boost voltage.

dc Abbreviation for direct current.

deflection The orderly movement of the electron beam in a picture (cathode-ray) tube. Horizontal deflection pertains to the left-right movement, vertical deflection, and the up-down movement of the beam.

deflection IC Today, the deflection circuits might have both the vertical and horizontal oscillator and amplifier circuits in one IC. You might find the deflection circuits in one large IC with many different circuits.

degaussing Demagnetizing. In color TVs, an internal or external circuit device that prevents or corrects any stray magnetization of the iron in the picture tube faceplate structure. Magnetization results in color distortion.

demodulator A demodulator separates or extracts the desired signal, such as sound energy or picture information, from its carrier.

detector Same as demodulator.

digital multimeter (DMM) The digital multimeter will measure voltage, resistance, current, and test diodes. Most have an LCD display. Some DMMs also measure capacitance, frequency, and tests transistors besides the regular DMM testing features.

diode A two-element electron device—a tube or semiconductor. The simplest and most common application of a diode is in the conversion of ac to dc (rectification).

discriminator An audio detector in an FM receiver or TV sound circuits. Also, a detector performing a similar function in other frequency-control circuits, such as horizontal frequency control.

electrolytic capacitor These capacitors can be used as filter or decoupling capacitors in the TV chassis. Large filter capacitors are used in the low-voltage power supply.

faceplate The front assembly of a picture tube. In a color tube, it includes the tri-color phosphor and the aperture mask.

field One scanning of the scene on the face of the picture tube, in which every alternate line is (temporarily) left blank. The scan duration of a field is $\frac{1}{60}$ second. Two fields, the second one filling in the blank lines left by the first one, make up a frame (a complete picture). A frame duration is $\frac{1}{30}$ second.

filter The electrolytic filter capacitor is located in the low-voltage power supply. Always replace it with one that has the same voltage and capacitance or higher (never lower values).

flyback, retrace A name given to return movement of the electron beam in a picture tube after completing each line and each field. You don't see flyback or retrace lines (normally) on the picture tube because they are blanked out.

flyback transformer Another name for the horizontal output transformer. The flyback transformer takes the sweep signal from the horizontal output transistor and builds up the high voltage to be rectified for the HV of the CRT. The flyback provides horizontal sweep for the yoke circuits.

focus Some picture tubes are constructed internally with self-focusing elements. in other TVs, a focus control varies the voltage applied to the picture-tube focus element. This voltage can vary from 4 to 5.3 kV, or 6kV to 9kV.

frame   The combination of two interlaced fields is called a *frame*. Because it consists of two fields each of 1/60 second duration, the frame duration is 1/30 second.

frequency   The number of recurring alternations in an electrical wave, such as ac, radio waves, etc. The frequency is specified by the number of alternations occurring during 1 second and given in hertz (cycles per second), kilohertz (1000 cycles), and megahertz (million cycles).

frequency counter   The frequency counter counts the frequency of various circuits. The frequency range could be from 2 Hz up to 100 MHz.

gain   Relative amplification. The number of times that a signal increases in size (level) caused by the action of one or more amplifiers. The overall gain of a signal often is millions of times.

gas   Refers to the presence (undesirable) of a trace of gas inside a vacuum tube. A gassy tube is a defective tube.

ghost   Most commonly a double-exposure type of a scene on the TV screen. Usually, a fainter picture appears somewhat offset to the right of the main image caused by the reception of two signals from the same station; one signal is delayed in time.

G-Y   The green color signal minus the monochrome.

high voltage   It generally refers to the multithousand picture tube voltage, but it can be used to mean any potential of a few hundred volts or more.

high-voltage probe   A test instrument that will check the anode and focus high voltage at the CRT. The new probes can measure up to 42,000 Vdc.

hold-down capacitors   The hold-down or tuning capacitors are used from collector terminal of horizontal output transistor to common ground. If the hold-down capacitor opens or dries up, the high voltage will increase, causing HV shutdown in the latest TVs.

horizontal   Pertaining to any of the functions associated with left-to-right scanning in a picture tube, including the horizontal amplifier, oscillator, frequency, drive, lock, AFC, etc.

H.O.T.   Horizontal output transformer, which steps up the low oscillator voltage with usually a driver and horizontal output transistor between. This voltage is rectified by the HV rectifier and is applied to the anode terminal of the picture tube.

hue   In color TV, the basic color characteristic that distinguishes red from green from blue, etc.

hum   Same as ac hum.

IC   Integrated circuit. A structure similar to a module, in which a number of parts required for the performance of a complete function are prewired and sealed. It is not repairable.

IF   Intermediate frequency. In the tuner of a TV or radio receiver, the incoming from the desired station is mixed with a locally generated signal to produce an intermediate signal, usually lower than the frequency of the incoming signal. The IF is the same for all stations. The tuner changes to accommodate each incoming signal.

IHVT   The integrated horizontal or high-voltage output transformer has HV diodes and capacitors molded inside of the flyback winding area. The new IHVT transformers can also provide several different voltage sources for the TV circuits.

in-line picture tubes A more recent development in color tube structure that produces the three basic colors in adjacent strips or bars, instead of the earlier types, which produced three-dot or triad groups. Improved color quality (as well as simplified design and maintenance) is claimed for this type of design.

isolation transformer The isolation transformer can be a variable type that raises or lowers the power-line voltage to the TV. Always use an isolation transformer with ac/dc TV.

intercarrier A term that describes the current system of TV receiver design, in which a common IF system is used both for picture and sound information. Older TVs implemented the split-sound design, in which separate IF channels for the picture and sound were used.

ion trap See trap.

kinnie The name often referred to as the *picture tube*.

leakage Undesired current flow through a component.

linearity Picture symmetry. Horizontal linearity pertains to symmetry between the right and left sides of the picture, best observed with a standard test pattern. Also, an adjustment for achieving such linearity. *Vertical linearity* refers to symmetry between the upper and lower halves of a picture.

line filter A device sometimes used between the ac wall outlet and a radio or TV to reduce or eliminate electrical noises.

line, transmission The antenna lead-in wire or cable.

lock, horizontal An adjustment in some TVs for setting the automatic frequency operation on the horizontal sweep oscillator.

loss Usually refers to the amount of signal loss in the antenna transmission line. This is particularly serious at UHF frequencies.

low-voltage regulator The low-voltage regulator is used in the low-voltage power supply. The regulator could be a transistor or located in an IC. The fixed regulator supplies a well-filtered, regulated, constant voltage.

microcompressor The microcompressor or microcomputer chip is built like a regular IC with 8 to 80 (or more) separate terminals. The microcompressor IC might have surface-mounted terminals.

modulation The process of combining (by superimposition) a sound or picture signal with a carrier signal for efficient transmission. The carrier's only function is to piggyback the intelligence.

module A subassembly of a number of parts, usually including transistors and diodes. It is encapsulated and not repairable. See IC.

modular chassis A TV chassis that is made up entirely of separate modules for each circuit in the TV.

motor boating A "putt-putt" sound caused in the audio sound input and output circuits. Motor-boating can be caused by poorly grounded or poorly filtered circuits.

oscillator Generator of a signal, such as the 3.58-MHz color subcarrier signal, the RF oscillator in the tuner, the horizontal oscillator signal (15,750 hertz), and the vertical oscillator signal (60 hertz).

oscilloscope A test instrument that can show exact waveforms throughout the TV circuits to help in troubleshooting and locating defective components.

PC board   Printed circuit board. A subassembly of various parts, not necessarily all for one and the same function, on a phenolic or fiberglass board on which the interconnections are printed on metal veins or paths. No conventional wiring, except external interconnections, is used.

parallel   A method of circuit component connection, where all components involved connect to common points so that each component is independent of all other components. For example, all light bulbs in your house are connected in parallel.

phosphor   The coating on the interior of the faceplate of a picture tube, which emits light when struck by an electron beam. The chemical composition of the phosphor determines the color of the light it will emit.

picture projection   Three small projection color tubes are used to project the TV image on the front or rear of a large-screen TV. The projection tubes are inside the cabinet of a rear-projection color set.

picture tube   The picture tube receives the video color signal that displays the picture on the picture-tube raster.

power supply   That portion of electronic equipment that provides operating voltages for its tubes, transistors, etc.

preamplifier   A high-gain amplifier used to build up a signal so that it is strong enough to present to the normal level amplifiers, for example, an antenna preamplifier for fringe-area reception.

pulse   A single signal of very short duration used for timing and sync purposes. Sync pulses are the best example of this type of signal. Pulses occur in precisely measured bursts.

purity, color   The display of the various true colors without any accidental or unwanted contamination of one color by any of the others. Color purity largely depends on correct convergence adjustments.

raster   The illuminated picture-tube screen fully scanned with or without video.

regulator   A transistor or IC that regulates the voltage for a given circuit in the low-voltage or HV power supplies.

remote control   A hand-held transmitter that controls the function of the TV by the operator from a distance.

resistance   Electrical friction represented by the letter R. The ohm (Ω) is the unit of resistance. Resistance limits current flow.

RF   Abbreviation for radio frequency.

retrace   The return movement of the scanning electron beam from the extreme right to the extreme left and from the bottom to the top of the raster. Also see flyback.

retrace blanking   The extinction or darkening of the light on the face of the picture tube during retrace time to make these lines invisible. If the retrace blanking fails, white lines (sloping downward from right to left) appear on the screen.

R-Y   The red color component of the overall color picture signal minus the monochrome.

sand castle   The sand-castle generator is a three-level signal pulse that includes horizontal and vertical blanking and burst keying pulses.

saturation   Pertains to the full depth of a color, in contrast to a faint, feeble color. Saturated colors are strong colors.

scanning lines   The horizontal lines that you can see up close in the picture or raster. The scanning lines make up the picture from left to right, looking at the front of the TV screen.

SCR   The silicon-controlled rectifier is used in the low- and high-voltage regulator power-supply circuits. In some TVs, an SCR is used as the horizontal output transistor.

semiconductor   A general name given to transistors, diodes, and similar devices in differentiation from vacuum-tube devices.

series   A connection between a number of components or tubes in chain fashion (i.e., one component follows the other). If any one component opens or burns out, it breaks the series circuit.

shadow mask   Same as aperture mask.

shield   A metallic enclosure or container that surrounds a component. Also see *tube shield*.

shielded cable   A wire with a metal casing on the outside to prevent unwanted electrical energy from reaching the inner conductor.

signal   Electrical energy containing intelligence (such as speech, music, pictures, etc.).

signal-to-noise ratio   A mathematic expression that indicates the relative strength of a signal within its noise environment. A good signal has a high signal-to-noise ratio.

solid-state   A term indicating that the radio, TV, etc., uses semiconductors (ICs, transistors, diodes, etc.), not vacuum tubes.

sound bars   Thick horizontal lines or bars, usually alternately dark and light, appearing on the TV picture screen caused by unwanted sound energy reaching the picture tube. In appearance, the width, number, and position of these bars varies with the nature of the sound. Sound bars are caused by a misadjusted circuit.

subcarrier   The color picture information carrier. It is called a *subcarrier* because it is a secondary carrier in the particular channel. The color subcarrier frequency is 3.58 MHz.

surface-mounted components   The surface-mounted parts are soldered into the circuit on the same side as the PC wiring. You might find surface-mounted components mounted under the PC chassis with larger components on top of the latest TVs.

sync   An abbreviation for a synchronizing signal. It is a timing signal or series of pulses sent by the transmitter and used by the receiver to stay in precise step with the transmitter.

sync clipper   See *clipper.*

sync separator   A circuit in a TV that separates the sync from the picture information or the vertical sync pulses from the horizontal sync pulses.

sweep marker generator   A generator used by the electronic technician for TV alignment.

transistor   A solid-state semiconductor used in the amplifier, oscillator, and power circuits of a TV. The transistor operates at lower voltage than the vacuum tube. Some chassis have both npn and pnp transistors.

trap   An electrical circuit that absorbs or contains a particular electrical signal (also called *wave trap*). Also a magnet, called an *ion trap*, used on the neck of some picture tubes for electron beam deflection.

triad   The three-color, three-dot group (red, green, and blue) of which the color picture tube phosphor is made. Each group of three dots is a triad, and there are thousands of triads on a modern color tube screen.

triac   A solid-state controller device usually located in the low-voltage power-supply circuits.

tripler   A solid-state component consisting of capacitors and diodes to triple the applied RF voltage from the flyback or horizontal output transformer. In the latest TVs, the horizontal output transformer and the high-voltage rectifiers can be molded into one component.

tube shield   A metal sleeve that fits snugly over a glass tube and shields it from extraneous electrical impulses. A tube shield is part of the tube's electrical circuit.

tuner   The tuner picks up each broadcast TV signal and passes it to the IF circuits for amplification. The tuner can be operated manually or with a remote control.

UHF   Ultrahigh frequencies, from 300 MHz upward. Channels 14 through 83 are all located in the UHF band and are, therefore, called *UHF stations*.

varactor   A semiconductor device with the characteristics of a tunable device through the application of a voltage. In contrast to the conventional frequency variation through the use of coil and capacitor techniques, the varactor requires only a voltage variation to affect tuning. In some recent TVs, varactor tuners have been used to replace the conventional coil-switching tuners. Simplicity, greater stability, and freedom from deterioration are claimed for this type of tuner.

variac   The variac can raise or lower the power-line voltage, which is helpful when trying to locate intermittents and defective flyback transformers.

vertical   Pertaining to the circuits and functions associated with the up-down motion or deflection of the electron beam.

vertical amplifier   An amplifier following the vertical oscillator used to enlarge the vertical sweep signal.

VHF   Very high frequencies, from 50 to 300 MHz. TV channels 2 through 13, as well as the FM band, are in the VHF spectrum.

video   A term applied to picture signals or information (video, circuits, video amplifier, etc.).

VIR   Vertical interval reference. Sometimes called *broadcast-controlled color-correction system*. It is an automatic, station-controlled signal used to initiate color- and luminance-corrective action in the TV, if so equipped. It is more elaborate and more effective than the automatic color control. See ACC.

VOM (volt-ohmmeter)   The first pocket-sized VOMs were used for continuity, voltage, resistance, and current tests. The VOM utilizes a meter to display the measured readings.

wave   The name given to each recurring variation in alternating electric energy, including radio and TV signals.

width   The width of a TV screen can be pulled in at each side, indicating problems within the horizontal deflection system. Poor width can be caused by the HV regulator transistors, SCRs, and zener diodes in the regulator circuit. Poorly soldered pincushion transformer connections can cause width problems.

X demodulator The designation of the red (R-Y) signal demodulator.

yoke Deflection yoke. The electrical assembly, somewhat in the shape of a yoke or collar, mounted on the picture tube neck against the flaring bell of the tube. The yoke imparts to the electron beam the scanning (left to right and top to bottom) or deflection to produce the raster and the image.

Y signal The picture-only (minus color) signal that is fed to the color picture tube. It is sometimes called the *brightness signal*, meaning the actual brightness and darkness (and all shades in between) of the picture. This brightness signal plus the red signal produce all the red hues in the picture. The same Y signal and blue signal produce all the blue coloration. The Y signal plus the green signal produce the green coloration to the scene.

Z demodulator Same as the X demodulator, except that it is for the blue (B-Y) signal.

zener diode A special semiconductor diode with a reverse breakdown voltage. The zener diode provides a fixed voltage in power circuits.

Index

About the Author

Homer L. Davidson has written more than 35 books and more than 1,000 articles in the field of technician-level electronics troubleshooting and repair. His highly popular books include *Troubleshooting and Repairing Audio Equipment,* Third Edition; *Troubleshooting and Repairing Compact Disk Players,* Third Edition; *Troubleshooting and Repairing Camcorders,* Second Edition; and *Troubleshooting and Repairing Solid-State TVs,* Third Edition. He is currently the TV Servicing Consultant for *Electronic Servicing & Technology* magazine.